Ammonia Oxidizing Bacteria
Applications in Industrial Wastewater Treatment

Ammonia Oxidizing Bacteria
Applications in Industrial Wastewater Treatment

Edited by

Maulin P. Shah
Enviro Technology Limited, India
Email: maulinshah1979@gmail.com

ROYAL SOCIETY OF **CHEMISTRY**

Print ISBN: 978-1-83916-687-7
PDF ISBN: 978-1-83767-196-0
EPUB ISBN: 978-1-83767-197-7

A catalogue record for this book is available from the British Library

The Royal Society of Chemistry is a charity, registered in England and Wales, Number 207890, and a company incorporated in England by Royal Charter (Registered No. RC000524), registered office: Burlington House, Piccadilly, London W1J 0BA, UK, Telephone: +44 (0)20 7437 8656.

Visit our website at www.rsc.org/books

Printed in the United Kingdom by CPI Group (UK) Ltd, Croydon, CR0 4YY, UK

Preface

At present, the release of ammonia in wastewater is an important environmental issue. Indeed, this phenomenon causes the eutrophication of water bodies and the subsequent depletion of dissolved oxygen. Further problems occur because certain forms of nitrogen (ammonia, nitrite, and nitrate) are highly toxic to aquatic life or may cause several issues following contamination with these compounds. Several biological and physical–chemical techniques for the removal of ammoniacal nitrogen compounds are available. Different nitrogen forms can be biologically removed by means of conventional activated sludge processes. Among the chemical-physical technologies, air stripping and breakpoint chlorination have been largely applied for the removal of ammoniacal nitrogen. Among the different solutions, microbiological or microbe-based approaches are very attractive because they promote the development of sustainable technologies. In this regard, Ammonia Oxidizing Bacteria have attracted great interest, specifically in wastewater treatment plants for the removal of ammoniacal nitrogen in recent years. Regarding the above, all of the chapters in this book will have high value in the area of wastewater treatment/wastewater microbiology.

Environmental eutrophication and/or pollution is a very serious concern worldwide. There is public outcry against the continually rising problem of pollution to ensure the safest and most eco-friendly environment for living beings. The removal of ammonia nitrogen through an active approach of using ammonia oxidizing bacteria has emerged as a sustainable and green solution of immense potential and promise. The microbial ecology, diversity and function of ammonia oxidizing bacteria in wastewater treatment has received a lot of attention in recent years due to its relatively low capital cost, eco-friendliness and high efficiency when compared with conventional clean-up technologies.

Maulin P. Shah

Ammonia Oxidizing Bacteria: Applications in Industrial Wastewater Treatment
Edited by Maulin P. Shah

Published by the Royal Society of Chemistry, www.rsc.org

Contents

Ammonia Oxidizing Bacteria: Applications in Industrial Wastewater Treatment
Edited by Maulin P. Shah

Published by the Royal Society of Chemistry, www.rsc.org

CHAPTER 1

Environmental Omics: A New Era to Study Ammonia-oxidizing Bacteria and Their Application in Bioremediation in the 21st Century

HIREN K. PATEL,*[a] PRIYANKA D. SHELADIYA[a] AND MAULIN P. SHAH[b]

[a] School of Science, P. P. Savani University, Surat, 394125, Gujarat, India; [b] Industrial Wastewater Research Lab, Enviro Technology Limited, Ankleshwar, Gujarat, India
*Email: drhkpatel1@gmail.com

1.1 Introduction

Nitrification is a microbial technique that converts ammonia to nitrate, and it is one of the key steps of the global nitrogen cycle, alongside nitrogen absorption and denitrification.[1] Nitrification begins with the oxidation of ammonia by a group of microscopic organisms known as 'AOBs' and archaea known as 'AOA.' The only byproduct of ammonia oxidizing bacteria, which were initially identified 125 years ago, is nitrite, which is then used as a free substrate by other microorganisms to generate nitrate. The first reports of 'comammox,' a bacteria that completes the oxidation of ammonia from ammonia to nitrate, were released in 2015.[2] This review has covered the enzymology, rules, and specific locations of AOB, AOA, and comammox in both natural and designed situations. Information on the routes and abiotic

Ammonia Oxidizing Bacteria: Applications in Industrial Wastewater Treatment
Edited by Maulin P. Shah

Published by the Royal Society of Chemistry, www.rsc.org

measures that cause N_2O is presented to explain the link of ammonia oxidizers with the ozone-harming chemical, nitrous oxide (N_2O).[2]

1.2 Chemolitho-autotrophic Ammonia Oxidation

Chemolithotrophic nitrification is a two-step process that converts ammonia to nitrite, which is then converted back to nitrate. There are no known autotrophic microscopic organisms that can catalyze the conversion of ammonia to nitrate. AOBs are chemolitho-autotrophs that require oxygen, although some species may be remarkably tolerant of low oxygen or anoxic circumstances.[3] In addition to their physiological consistency, Head *et al.*[4] used 16S rRNA quality grouping data to determine that all common AOB isolates are limited to two developmentally distinct ancestries of the Proteobacteria class. Similarly, all strains isolated from local and freshwater environments were assigned to a single (monophyletic) developmental group within a subclass of the class Proteobacteria and their ancestry was traced back to a single chemolithoautotrophic ammonia oxidizing progenitor. The exclusive conversion of ammonia to nitrite by AOBs involves a series of events linked by hydroxylamine, as shown in the following equations.

$$2H^+ + NH_3 + 2e^- + O_2 \rightarrow NH_2OH + H_2O \tag{1.1}$$

$$NH_2OH + H_2O \rightarrow HONO + 4e^- + 4H^+. \tag{1.2}$$

It is generally accepted that ammonia (NH_3) rather than ammonium (NH_4^+) is used as a substrate, and hence the amount of ammonia/ammonium may interfere with the reaction.[5] The main reaction is catalyzed by a layer-bound multisubunit complex called ammonia monooxygenase (AMO), while the second is catalyzed by a periplasm-related enzyme called hydroxylamine oxidoreductase (HAO). Two of the electrons produced in the subsequent reaction are used to replace the initial reaction's electron commitment, while the other two are transferred to the terminal oxidase *via* an electron transfer chain, resulting in a proton motive force.

$$2H^+ + \frac{1}{2}O + 2e^- \rightarrow H_2O. \tag{1.3}$$

Eqn (1.4) gives the overall equation:

$$NH_3 + 1\frac{1}{2}O_2 \rightarrow HONO + H_2O. \tag{1.4}$$

Despite their incapability to be cultivated on carbon-based substrates, the hydrolysis of urea (eqn (1.5)) can be used by some species in place of a primary ammonia source.

$$(NH_2)_2C{=}O + H_2O \rightarrow 2NH_3 + CO_2. \tag{1.5}$$

In any case, the anaerobic oxidation of ammonia has two possibilities. The first was discovered using common oxygen-consuming ammonia oxidizing strains that may reduce nitrate or nitrogen dioxide by using hydroxylamine or ammonia instead of electron contributors.[6] The Anammox technique, a second anaerobic-autotrophic ammonia oxidation method, has been credited to a group of life forms inside a deep-reaching clade of Planctomycetales.[7,8] The devices mastered by these anaerobes produce N_2 from ammonia and nitrate from hydroxylamine and hydrazine as intermediates.[9] It is unknown how much anaerobic ammonia oxidation occurs in native settings.

1.3 Role of Ammonia-oxidizing Bacteria in the Nitrogen Cycle

The most common natural sources of ecological ammonia are nitrogen fixation and mineralization, with the latter being most directly linked to autotrophic ammonia oxidation. Overall, the amount of ammonia used, either directly or indirectly, for farming and modern applications has greatly increased in recent times.[10–12] The oxidation of ammonia to nitrate is the major phase in the oxidation of ammonia to nitrate, and so dominates the entire nitrogen cycle. In many reaction designs, ammonia oxidation is regarded to be the rate-limiting step for nitrification; for example, nitrite is rarely produced such that it collects in the environment.[13–16] Through at least three different mechanisms, nitrification can contribute to a net loss of nitrogen from ecological systems. To begin with, the final production of nitrate can be used to power denitrification, which is the primary method for safely removing nitrogen from ecological systems. Second, the product of ammonia oxidation, nitrite, can undergo chemodenitrification; however, it's unclear to what extent this process affects nitrogen imbalances. Finally, nitrate spillage into the soil is a third key source of nitrogen disaster. While anionic NH_4^+ particles bind strongly to cationic substances in the soil, NO_3 is far more mobile and hence spills quickly into neighboring channels.

1.4 Isolation and Cultivation of New Ammonia-oxidizers

Our knowledge of ammonia oxidizers and the challenges of their physiology has been greatly altered through examining axenic societies. The main seclusion of a ammonia oxidizer, *Nitrosomonas europaea*, was described in 1890 by Sergei.[17] Since then, a variety of strains of AOB, AOA, and comammox microbes have been cultured from a variety of biological systems, including soils, estuaries, marine environments, underground aquifers, freshwater, residues, aquaria, wastewater treatment facilities, heated water pipe biofilms, drinking water plants, and various others.[18–20] Ammonia salt oxidizers resist adhering to the close-knit networks used by microorganisms to eliminate their metabolic byproducts, protect them from oxidative stressors, or achieve mutual coordination of activities, such as cyanate oxidation to ammonia salts through related nitrite oxidizers.[21,22] This means that their separation is far from an insignificant task. As a result, ammonia oxidizers are extremely vulnerable to losing their catalytic activity. Initially thought to resist mixotrophic digestion, the involvement of pyruvate in the growth phase of AOA was later discovered to detoxify hydrogen peroxide.[23]

1.5 Commercial Significance and Application of Ammonia-oxidizing Bacteria

The key stage in the disposal of nitrogen throughout waste activity is the oxidation of organic ammonia. The treatment of nitrogenous waste has become a critical issue in the environment as a result of the rise in nitrogenous waste due to the improvement of animal farming, human activities, and nitrogen-producing industries. The removal of nitrogen from waste is extremely important from an ecological standpoint, as the release of untreated waste can result in extreme eutrophication of the environment, particularly in heavily populated areas. In fact, nitrification helps to prevent environmental contamination with possibly toxic ammonia salts, in spite of the fact that the system doesn't need it for successful denitrification. Waste treatment is by far the most important biotechnological application of AOBs. Because of the broad substrate scope of the ammonia monooxygenase compound, which is found in all AOBs, various disordered aliphatic, aromatic, and halogenated particles are frequently co-oxidized.[24] AOBs may function in bioremediation systems that include a preliminary oxidation step, which is typically the rate-restricting adjustment in the microbial sludge of many active hydrocarbons. Although preliminary evidence from unadulterated culture samples suggests that AOBs may play a role in methane oxidation,[25] enabling the removal of this key ozone-depleting component, field findings have yet to demonstrate significant methane uptake through AOBs.[26,27]

AOBs could also play a role in biofiltration systems, which typically include microbial communities that utilize influent substrates to maintain relatively constant biomass and output waste that is free of input debasements. Biofilter plants have been used for a variety of applications to date, most notably the elimination of odors associated with waste treatment and soil fertilization, but also for more unusual applications such as determining whether long-distance sifting estimations are reasonable for operated spacelabs.[28]

1.6 Genomics and Metabolic Models

The reporting of potential microbial species has been identified through genome sequencing. Genome sequencing provides a platform for extrapolating genome-scale metabolic models which also allows the differentiation of the physiological, regulating, and transformational properties of microorganisms. With regard to the acid-resilience of the AOA '*Candidatus Nitrosotalea devanaterra*'[29] and related strains, genomics offers potential explanations for certain advantageous physiologies. Differentiation of four '*Ca. Nitrosotalea's* genome analyses with 23 extra AOA genomes uncovered 743 shared center proteins,[30] which is less than the 860 mutual center proteins depicted in an earlier AOA genome comparison.[31] All AOA genomes encode the 3-hydroxypropionate/4-hydroxybutyrate CO_2 fixation pathway, predominant carbon pathways, and proteins for ammonia oxidation. A Na^+/solute symporter, metal carriers, and a chaperone specialised for proline-rich proteins are also found in the genomes of certain '*Ca. Nitrosotalea*'.[30] The genomes of soil AOA encode exclusive highlights; for example, envelope adjustments for biofilm creation, polysaccharide production, and cell–cell grip.[31] Earlier analyses of the '*Ca. N. devanaterra*' genome identified 51 distinct candidate genes for low pH resistance with high-affinity substrate transport, film impermeability, and various cation carriers;[29] however, a more thorough investigation revealed that all but 10 of these genes are in the AOA genomes of non-acidophilus species.[30] These studies feature the basics for partnering throughout as many genomic groups as permitted to characterise distinct quality sets that address unique capacities, but they also suggest that quality rundown alone is a flawed indication of particular pathways and capacities. It is necessary to gather many levels of useful information from a microbe and express it numerically in order to reveal the genome-scale metabolic structure. The model can precisely predict universal metabolic information about an organism before validation through wet-lab testing when repeatedly complexed with trial data, yielding useful data far beyond genomic grouping alone. Metabolic organization models have been utilized to anticipate N_2O improvements from nitric oxide (NO)-delivering reactions in AOB consortia through nitrite-oxidizing microorganisms[32,33] and to characterize the anoxic–oxic transformation of *N. europaea*.[34] According to the most recent study, repeated anoxic–oxic cycles in *N. europaea* led to a decrease in N_2O production and progressive alterations in the transition of nitrogen oxide metabolites, which were connected to modifications at the protein level, perhaps more so than transcriptional changes.[34]

1.7 Introduction to Environmental Proteomics

A while back, Wilmes and Bond[35] characterized metaproteomics as "the significant portrayal of the entire protein supplement of the natural microbiota at a predefined point in a period"; subsequently, the rapid and multifold acquisition of high-throughput "omics" information has resulted in the definition of numerous new fields, including ecological proteomics, metaproteomics, local area proteomics, and local area proteogenomics. Natural proteomics is a conventional term essentially

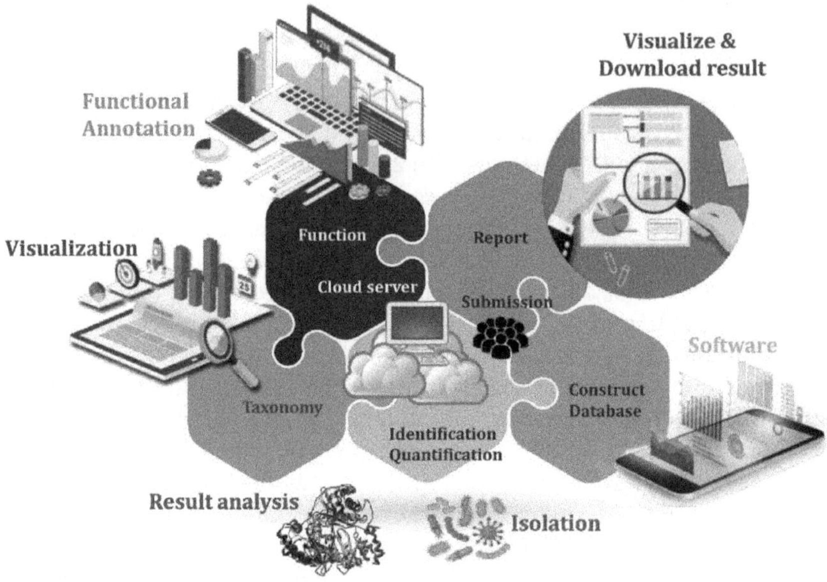

Figure 1.1 Overview of the functional process of environmental omics.

depicting the proteome examination of ecological examples; metaproteomics incorporates investigations of incredibly complex organic plans, and doesn't allow the transmission of the enormous number of proteins to the correct species inside the phylotypes (Figure 1.1). In contrast, in local area proteomics, the majority of the perceived proteins can be related to specific individuals in the local area; thus far, such investigations have been under low- or medium-intricacy conditions. The word proteogenomics, which is used to characterize the use of proteomics for the enhancement of quality comments, does as of now additionally depict the evaluation of the strain or species varieties and the transformative improvement of the genomics cosmetics to specific conditions.[36] Moreover, proteogenomics adds insight into the genuine quality capacity by connecting perceived protein data to the DNA level.

1.8 Microbial Culture Proteomic Study Techniques

Much work has been devoted to investigating microbial strategies in recent years. Proteomic data collection and a few common breakthroughs have been established, including test selection, protein separation, protein partitioning and fractionation, mass spectrometry analysis, data set searching, and long-term data comprehension. Attributable to the different presence of ecological examples, when working with marine and freshwater just as surface examples,[37] remarkable techniques have been developed for either test handling or protein extraction. Blended proteins are typically segregated utilizing possibly one-dimensional or two-dimensional electrophoresis with polyacrylamide gel (2-D PAGE). The objective protein spots or groups are then extracted and the proteins are processed with trypsin or different chemicals into peptides. The subsequent peptides are then identified by mass spectrometry (MS) or tandem mass spectrometry (MS/MS), data set sweeps, and bioinformatic investigation.[38] The other method is fluid chromatography (LC)-based interaction, where the whole proteome is processed utilizing proteases without earlier gel protein partitioning into a more complicated peptide combination. Also, using strong cation exchange chromatography or microcapillary techniques, the resultant peptides are isolated. All in all, utilizing fluid chromatography combined with MS/MS (LC-MS/MS), the separated peptides are analyzed. For protein recognition and subsequent bioinformatics examination, the produced MS information is deciphered. This strategy overcomes the deficiencies of gel-based methodologies and significantly improves proteome inclusion compared with the gel-based methodology, enabling a large number of proteins to be recognized in a high-throughput manner inside a restricted timeframe[36] and in turn, making it conceivable to distinguish insoluble film proteins.[39] Subsequently, this LC-based strategy has become the standard of proteomic studies in the microbial world, even though it experiences problems with reproducibility, dynamic range, and shortage of data sets. The repeatability of metaproteomic analyses is important in determining whether the change in protein expression in the microbial community is inherent or the result of experimental error. Specialized reproducibility will generally be near half across sets of three and over 67% across repetitions using a similar MS stage.[40]

1.9 Potential Applications of Environmental Proteomics

In their ordinary living space, organisms are often confronted with fast and correctional changes of natural cut-off points, like temperature, moistness, predators, and supplement accessibility. A shared methodology of microorganisms to overcome these variations is a change of their protein expression profiles. Likewise, the basic investigation of the different qualities and their guidelines isn't sufficient to completely understand microbial transformation approaches, and post-genomics examinations with transcriptomics and proteomics are promptly expected for the examination of the physiology of the complex microbial relationship at a sub-atomic level. Ecological proteomics initially contains a genuine "money box" of information, ranging from the basic protein list to relative and quantitative proteomics, investigations of protein limitations, the identification of post-translational changes that may influence the usefulness of the proteins, investigation of the protein–protein associations, and the assurance of the amino acid sequences and their genotypes. Consequently, most likely the use of the higher up advances in microbial nature will be bountiful and contain a record of new useful qualities, the documentation of new synergistic catalysts or complete metabolic pathways, and the record of a practical bioindicator to screen elements and the maintainability of the ecological class.[41]

1.10 Enzymology of Ammonia-oxidation

AOBs, AOA, and comammox microbes oxidize ammonia to hydroxylamine utilizing protected ammonia monooxygenase (AMO) chemicals.[42,43] All ammonia oxidizers correspondingly oxidize hydroxylamine and yield nitrite; however, the enzymology for this piece of the pathway remains partly uncertain. In 2017, Caranto *et al.*[44] found that the chemical hydroxylamine dehydrogenase (HAO) of AOBs oxidizes hydroxylamine to NO rather than nitrite.[45] This finding disproved the long-standing assumption of hydroxylamine being directly oxidised by HAO to nitrite and led to another test for nitric oxide oxidoreductase (NOO). In this manner, two NOO candidates have been suggested for AOBs: nitrosocyanin, encoded through ncyA, and the contrarily working copper-containing nitrite reductase, encoded through nirK.[46] All AOB genomes have the NCAA gene, which is unique to AOBs and is absent only in the genome of the oligotrophic strain *Nitrosomonas sp.* Is79.[47] A proteomic analysis of three AOB species revealed that NcyA was highly expressed with additional compounds in the ammonia oxidation pathway, demonstrating its crucial function.[18] Similarly, Nike's appearance in AOBs isn't constitutive and is in turn controlled through sensitivity to high-ammonia,[18] NO,[48] and anoxia. Nike is likewise absent from some AOB genomes.[49] Likewise, the nirK gene can be taken out in *N. europaea* with no detrimental result on development, confirming that it is non-essential in this model AOB strain.[50]

1.11 Ammonia-oxidizers in the Environment and Production of N_2O

The majority of cultivated AOBs encode and express NirK and nitric oxide reductase enzymes (NorB/Y), which in the model should reduce nitrite to NO and N_2O. But, the NorB enzyme was also found to be important to the nitrifier denitrification activity in a gene knock-out study of *N. europaea*.[50] Furthermore, oligotrophic strains that lack NorB cannot perform nitrifier denitrification and the strain *Nitrosomonas communis* lacks NirK, but hitherto retains nitrifier denitrification activity.[49] These studies specify that unidentified nitrite reductases are yet to be discovered in AOBs, and, meanwhile, the physiological role of NirK remains unknown. Other mechanisms of N_2O generation by AOBs include the anaerobic oxidation of hydroxylamine through the enzyme cytochrome P460[45] and abiotic reactions of metabolic intermediates, hydroxylamine, and nitrite by each auxiliary, with media components, and with metals.[49,51–54]

Interestingly, AOA does not have the physiological capacity to perform nitrifier denitrification.[49] Abiotic responses of metabolic intermediates harvest N_2O, which can be considered where AOA is rich and active, such as marine ecosystems.[55] Abiotic formation of N_2O from intermediates of AOA metabolism has been long-established in soils and is favored under conditions of low ammonium supply.[56] These studies have established that the manufacture of N_2O through nitrifier denitrification in AOBs is significantly better than abiotic N_2O manufacture through AOA and is additionally favored in soils with high ammonium supply.[57,58] Studies are ongoing to define whether comammox bacteria contribute to N_2O emissions; however, their strict oligotrophic lifestyle would designate that they produce very little, if any. It should be noted, however, that clarification of the exact sources of N_2O is an ongoing methodological challenge, complicated by the lack of inter-laboratory calibration of standards and instrumentation.

1.12 Remediation of Recalcitrant Pollutants

Ammonia monooxygenase has a broad substrate range, which allows AOBs to grow on a variety of substrates, some of which can be incredibly plentiful and prevalent in the environment. Anthropogenic toxins oxidized through β-AOBs include trichloroethylene, chloroethane, benzene, various halogenated aliphatics, dimethyl ether, and characteristic and engineered estrogens.[19] While several of these substrates are typically thought of as inhibitors of autotrophic ammonia oxidation, AOBs don't need energy to oxidise them. Likewise, the redox environment can affect the bioavailability of numerous contaminants, such as heavy metals, as well as the movement of extra microorganisms that play huge roles in the remediation of such mixtures.[59] In this manner, the activity of AOBs may have both immediate and long-term effects in ecological cleanup applications. It is thought that monitoring the health and species strategy of β-AOBs in unclean environments and bioreactors constitutes a negligible portion of microbial biology. However, it ought to be noticed that lab-based examinations in this space have all decided on *N. europaea*, a nitrifier not bountiful in various natural settings, and that ecological mediums disturb the pace of co-oxidation of toxins.[59]

1.13 Conclusion

Our knowledge of ammonia oxidizing microorganisms has expanded in recent years through the discovery and development of novel organisms, the quick propagation of genome sequencing and useful 'omics information, concentrated biochemical work that both delayed the usefulness of recognized proteins and moved us to characterize novel chemicals, and environment examinations that uncovered interconnections of ammonia oxidizers by all other and by accomplice microorganisms. The presence of N_2O in the air is now supported by both enzymology and abiotic relationships, which exist between ammonia oxidizers and various other systems. We additionally recognize that NO is a fundamental intermediate in all ammonia oxidation pathways. Remaining inquiries include the identification of NOO proteins, the different roles of NirK and rubredoxin for NO, the capacity for comammox microorganisms to yield N_2O, and genomic-organization highlights that allow particular capacities and inclinations among AOBs, AOA, and comammox microscopic organisms. The aim of these investigations is to combine ammonia oxidizers for contemporary and ecological applications, as well as to truly mitigate the use of compost through N_2O production and ammonia oxidizer employment. Numerous revelations remain to be discovered as key highlights of these intriguing microorganisms and their selective digestion remains uncertain.

References

1. L. Y. Stein and M. G. Klotz, The nitrogen cycle, *Curr. Biol.*, 2016, **26**(3), R94–R98.
2. H. Daims, *et al.*, Complete nitrification by Nitrospira bacteria, *Nature*, 2015, **528**(7583), 504–509.
3. P. Bodelier, *et al.*, Dynamics of nitrification and denitrification in root-oxygenated sediments and adaptation of ammonia-oxidizing bacteria to low-oxygen or anoxic habitats, *Appl. Environ. Microbiol.*, 1996, **62**(11), 4100–4107.
4. I. M. Head, *et al.*, The phylogeny of autotrophic ammonia-oxidizing bacteria as determined by analysis of 16S ribosomal RNA gene sequences, *J. Gen. Microbiol.*, 1993, **139**(6), 1147–1153.
5. P. M. Wood, *Nitrification as a bacterial energy source*, Society for General Microbiology, Oxford, 1986, pp. 39–62.
6. I. Schmidt and E. Bock, Anaerobic ammonia oxidation with nitrogen dioxide by Nitrosomonas eutropha, *Arch. microbiol.*, 1997, **167**, 106–111.
7. M. Schmid, *et al.*, Molecular evidence for genus level diversity of bacteria capable of catalyzing anaerobic ammonium oxidation, *Syst. Appl. Microbiol.*, 2000, **23**(1), 93–106.
8. M. Strous, *et al.*, Missing lithotroph identified as new planctomycete, *Nature*, 1999, **400**(6743), 446–449.
9. A. Mulder, *et al.*, Anaerobic ammonium oxidation discovered in a denitrifying fluidized bed reactor, *FEMS Microbiol. Ecol.*, 1995, **16**(3), 177–183.
10. R. Bobbink, M. Hornung and J. G. M. Roelofs, The effects of air-borne nitrogen pollutants on species diversity in natural and semi-natural European vegetation, *J. Ecol.*, 1998, **86**(5), 717–738.
11. J. N. Galloway, Acid deposition: perspectives in time and space, *Water, Air, Soil Pollut.*, 1995, **85**, 15–24.
12. P. Grennfelt and H. Hultberg, Effects of nitrogen deposition on the acidification of terrestrial and aquatic ecosystems, *Water, Air, Soil Pollut.*, 1986, **30**, 945–963.

13. W. De Boer, *et al.*, Nitrification in Dutch heathland soils: II. Characteristics of nitrate production, *Plant Soil*, 1990, **127**, 193–200.
14. W. De Boer, *et al.*, The chemolithotrophic ammonium-oxidizing community in a nitrogen-saturated acid forest soil in relation to pH-dependent nitrifying activity, *Soil Biol. Biochem.*, 1992, **24**(3), 229–234.
15. M. E. El-Demerdash and J. C. G. Ottow, Einfluss einer hohen Nitratdüngung auf Kinetik und Gaszusammensetzung der Denitrifikation in unterschiedlichen Böden, *Zeitschrift für Pflanzenernährung und Bodenkunde*, 1983, **146**(2), 138–150.
16. J. J. Prosser, Autotrophic nitrification in bacteria, *Bull. de la Société d'anthropologie de Paris*, 1990, **30**, 125–181.
17. S. Winogradsky, Recherches sur les organisms de la nitrification, *Ann. Inst. Pasteur.*, 1890, **4**, 213–231.
18. L. Y. Stein, Insights into the physiology of ammonia-oxidizing microorganisms, *Curr. Opin. Chem. Biol.*, 2019, **49**, 9–15.
19. H. K. Patel, *et al.*, Omics: A revolutionary tool to study ammonia-oxidizing bacteria and their application in bioremediation, in *Development in Wastewater Treatment Research and Processes*, Elsevier, 2022, pp. 61–81.
20. K. D. Kits, *et al.*, Kinetic analysis of a complete nitrifier reveals an oligotrophic lifestyle, *Nature*, 2017, **549**(7671), 269–272.
21. C. J. Sedlacek, *et al.*, Effects of bacterial community members on the proteome of the ammonia-oxidizing bacterium Nitrosomonas sp. strain Is79, *Appl. Environ. Microbiol.*, 2016, **82**(15), 4776–4788.
22. M. Palatinszky, *et al.*, Cyanate as an energy source for nitrifiers, *Nature*, 2015, **524**(7563), 105–108.
23. J.-G. Kim, *et al.*, Hydrogen peroxide detoxification is a key mechanism for growth of ammonia-oxidizing archaea, *Proc. Natl. Acad. Sci.*, 2016, **113**(28), 7888–7893.
24. A. B. Hooper, *et al.*, Enzymology of the oxidation of ammonia to nitrite by bacteria, *Antonie van Leeuwenhoek*, 1997, **71**, 59–67.
25. P. Steudler, *et al.*, Microbial controls of methane oxidation in temperate forest and agricultural soils, in *Microbiology of Atmospheric Trace Gases: Sources, Sinks and Global Change Processes*, Springer, 1996.
26. P. L. E. Bodelier and P. Frenzel, Contribution of methanotrophic and nitrifying bacteria to CH_4 and NH_4^+ oxidation in the rhizosphere of rice plants as determined by new methods of discrimination, *Appl. Environ. Microbiol.*, 1999, **65**(5), 1826–1833.
27. P. L. Bodelier, *et al.*, Stimulation by ammonium-based fertilizers of methane oxidation in soil around rice roots, *Nature*, 2000, **403**(6768), 421–424.
28. J. A. Joshi, *et al.*, Biological removal of gaseous ammonia in biofilters: space travel and earth-based applications, *J. Air Waste Manage. Assoc.*, 2000, **50**(9), 1647–1654.
29. L. E. Lehtovirta-Morley, *et al.*, Identifying potential mechanisms enabling acidophily in the ammonia-oxidizing archaeon "Candidatus Nitrosotalea devanaterra", *Appl. Environ. Microbiol.*, 2016, **82**(9), 2608–2619.

30. C. W. Herbold, *et al.*, Ammonia-oxidising archaea living at low pH: insights from comparative genomics, *Environ. Microbiol.*, 2017, **19**(12), 4939–4952.
31. M. Kerou, *et al.*, Proteomics and comparative genomics of Nitrososphaera viennensis reveal the core genome and adaptations of archaeal ammonia oxidizers, *Proc. Natl. Acad. Sci.*, 2016, **113**(49), E7937–E7946.
32. O. Perez-Garcia, *et al.*, Assessment of nitric oxide (NO) redox reactions contribution to nitrous oxide (N_2O) formation during nitrification using a multispecies metabolic network model, *Biotechnol. Bioeng.*, 2016, **113**(5), 1124–1136.
33. B. L. Mellbye, *et al.*, Genome-scale, constraint-based modeling of nitrogen oxide fluxes during coculture of Nitrosomonas europaea and Nitrobacter winogradskyi, *mSystems*, 2018, **3**(3), e00170–17.
34. R. Yu, *et al.*, Nitrosomonas europaea adaptation to anoxic-oxic cycling: insights from transcription analysis, proteomics and metabolic network modeling, *Sci. Total Environ.*, 2018, **615**, 1566–1573.
35. P. Wilmes, *et al.*, Community proteogenomics highlights microbial strain-variant protein expression within activated sludge performing enhanced biological phosphorus removal, *ISME J*, 2008, **2**(8), 853–864.
36. N. C. VerBerkmoes, *et al.*, Functional analysis of natural microbial consortia using community proteomics, *Nat. Rev. Microbiol.*, 2009, **7**(3), 196–205.
37. K. S. Habicht, *et al.*, Comparative proteomics and activity of a green sulfur bacterium through the water column of Lake Cadagno, Switzerland, *Environ. Microbiol.*, 2011, **13**(1), 203–215.
38. D.-Z. Wang, Z.-X. Xie and S.-F. Zhang, Marine metaproteomics: current status and future directions, *J. Proteomics*, 2014, **97**, 27–35.
39. C. C. Wu and J. R. YatesIII, The application of mass spectrometry to membrane proteomics, *Nat Biotechnol.*, 2003, **21**(3), 262–267.
40. J. E. Elias, *et al.*, Comparative evaluation of mass spectrometry platforms used in large-scale proteomics investigations, *Nat. Methods*, 2005, **2**(9), 667–675.
41. P.-A. Maron, *et al.*, Metaproteomics: a new approach for studying functional microbial ecology, *Microb. Ecol.*, 2007, **53**, 486–493.
42. P. Pjevac, *et al.*, AmoA-targeted polymerase chain reaction primers for the specific detection and quantification of comammox Nitrospira in the environment, *Front Microbiol.*, 2017, **8**, 1508.
43. R. J. E. Alves, *et al.*, Unifying the global phylogeny and environmental distribution of ammonia-oxidising archaea based on amoA genes, *Nat. Commun.*, 2018, **9**(1), 1517.
44. J. D. Caranto and K. M. Lancaster, Nitric oxide is an obligate bacterial nitrification intermediate produced by hydroxylamine oxidoreductase, *Proc. Natl. Acad. Sci.*, 2017, **114**(31), 8217–8222.
45. J. D. Caranto, A. C. Vilbert and K. M. Lancaster, Nitrosomonas europaea cytochrome P460 is a direct link between nitrification and nitrous oxide emission, *Proc. Natl. Acad. Sci.*, 2016, **113**(51), 14704–14709.

46. K. M. Lancaster, *et al.*, Alternative bioenergy: updates to and challenges in nitrification metalloenzymology, *Joule*, 2018, **2**(3), 421–441.
47. A. Bollmann, *et al.*, Complete genome sequence of Nitrosomonas sp. Is79, an ammonia oxidizing bacterium adapted to low ammonium concentrations, *Appl. Environ. Microbiol.*, 2013, **7**, 469–482.
48. H. J. Beaumont, *et al.*, Expression of nitrite reductase in Nitrosomonas europaea involves NsrR, a novel nitrite-sensitive transcription repressor, *Mol. Microbiol.*, 2004, **54**(1), 148–158.
49. J. A. Kozlowski, *et al.*, Pathways and key intermediates required for obligate aerobic ammonia-dependent chemolithotrophy in bacteria and Thaumarchaeota, *ISME J.*, 2016, **10**(8), 1836–1845.
50. J. A. Kozlowski, *et al.*, Revision of N_2O-producing pathways in the ammonia-oxidizing bacterium Nitrosomonas europaea ATCC 19718, *Appl. Environ. Microbiol.*, 2014, **80**(16), 4930–4935.
51. S. Liu, *et al.*, Abiotic conversion of extracellular NH_2OH contributes to N_2O emission during ammonia oxidation, *Environ. Sci. Technol.*, 2017, **51**(22), 13122–13132.
52. A. Terada, *et al.*, Hybrid nitrous oxide production from a partial nitrifying bioreactor: hydroxylamine interactions with nitrite, *Environ. Sci. Technol.*, 2017, **51**(5), 2748–2756.
53. C. H. Frame, *et al.*, Acidification enhances hybrid N_2O production associated with aquatic ammonia-oxidizing microorganisms, *Front. Microbiol.*, 2017, **7**, 2104.
54. X. Zhu-Barker, *et al.*, The importance of abiotic reactions for nitrous oxide production, *Biogeochemistry*, 2015, **126**, 251–267.
55. M. Trimmer, *et al.*, Nitrous oxide as a function of oxygen and archaeal gene abundance in the North Pacific, *Nat. Commun.*, 2016, **7**(1), 13451.
56. L. Hink, *et al.*, The consequences of niche and physiological differentiation of archaeal and bacterial ammonia oxidisers for nitrous oxide emissions, *ISME J.*, 2018, **12**(4), 1084–1093.
57. L. Hink, *et al.*, Kinetics of NH_3-oxidation, NO-turnover, N_2O-production and electron flow during oxygen depletion in model bacterial and archaeal ammonia oxidisers, *Environ. Microbiol.*, 2017, **19**(12), 4882–4896.
58. L. Hink, G. W. Nicol and J. I. Prosser, Archaea produce lower yields of N_2O than bacteria during aerobic ammonia oxidation in soil, *Environ. Microbiol.*, 2017, **19**(12), 4829–4837.
59. L. A. Sayavedra-Soto, *et al.*, Nitrification and degradation of halogenated hydrocarbons—a tenuous balance for ammonia-oxidizing bacteria, *Appl. Microbiol. Biotechnol*, 2010, **86**, 435–444.

CHAPTER 2

Nitrification in the Ocean: A Systematic Review

DIPRO MUKHERJEE,[†,a] MOUPRIYA NAG,[†,b] DIBYAJIT LAHIRI,*[b] ANKITA DEY[c] AND RINA RANI RAY*[d]

[a] Department of Bioscience and Bioengineering, Indian Institute of Technology (IIT), Jodhpur, Rajasthan, India; [b] Department of Biotechnology, University of Engineering & Management, Kolkata, West Bengal, India; [c] Department of Microbiology, Belle Vue Clinic, Kolkata, West Bengal, India; [d] Department of Biotechnology, Maulana Abul Kalam Azad University of Technology, Haringhata, West Bengal, India

*Emails: dibyajit.lahiri@uem.edu.in; raypumicro@gmail.com

2.1 Introduction

Nitrogen acts as an important element for the maintenance of life and its functioning within marine ecosystems. It plays an important role in biogeochemistry.[1] The connection of nitrogen with various other elements involved in the cycle, such as carbon, means that it possesses a strong role in the climatic condition of the Earth. The oxidation state of nitrogen within the ocean varies from –III in its reduced state, like in the ammonium ion (NH_4^+), to +V within fully oxidized nitrates (NO_3^-).[2] These oxidation states help in understanding the importance of nitrogen both as an electron acceptor and electron donor during the process of energy metabolism in the marine ecosystem. Microorganisms help in mediating this redox transformation with a change in the concentration of nitrogenous compounds within the environment.[3] The availability of nitrogen helps in the regulation

[†]Equal contribution.

Ammonia Oxidizing Bacteria: Applications in Industrial Wastewater Treatment
Edited by Maulin P. Shah

Published by the Royal Society of Chemistry, www.rsc.org

of the activity of the biological pumps resulting in the uptake of atmospheric carbon dioxide. The major source of nitrogen observed within the ocean is by the process of biological nitrogen fixation, whereas the major sinks comprise the processes ammonium oxidation and denitrification.[4] Because alterations of this balance caused by anthropogenic activity may have a significant impact on the health of marine ecosystems, biodiversity and climate change, the study of microbial communities involved in marine N cycling has attracted great interest in recent years.[5]

Dissolved inorganic nitrogen concentration is an important component among every form of life. However, deep inside the ocean, nitrogen majorly exists in the form of inert gas N_2, more than ninety percent of which again exists in the dissolved state. This percentage cannot be accessed by almost all species.[6] Other the inert nitrogen gas, nitrogen exists as reactive nitrogen, which comprises mainly ammonia, nitrate and other organic compounds of nitrogen that exist in their dissolved forms. The nitrate concentration in the water near the surface is almost equal to zero. The nitrate concentration is also very low in the tropical and subtropical regions. But in the temperate water bodies and in the Arctic and Antarctic oceans, the nitrate concentration is very high and thousands of tons are present in micro-molar concentrations.[7] The rise in the level of nutritional elements in the deepest parts of the ocean is the reason for a circular pattern occurring globally, which is known as the "oceanic conveyor belt". It extends from the northern Atlantic and continues up to the Pacific and Indian oceans. One of the reasons for the existence of this oceanic conveyor belt is the presence of excess amounts of salt, which drives this belt.[8] Mostly, this is evident in the northern Atlantic Ocean, in which we can see that the water on the surface sometimes goes to the bottom of the ocean and travels more in the southern direction. A part of the water current located towards the south maintains a steady supply of water towards the Antarctic Circumpolar Current, which again continues flowing towards the Indian and Pacific oceans.[9] Again, from the Indian and Pacific oceans, the water flows back towards the Atlantic Ocean, flowing through the Tasman Sea waters and the waters of the Indonesian islands. The water currents reside in the Atlantic water for over a hundred and eighty years and then the water returns into the deep ocean, which may take up to thousands of years.[10]

There are four steps or processes that circulate the water carrying the nutritional elements to the surface area. These are winter mixing, diffusion, upwelling and eddies.[11] All of these are operated through a seasonal thermocline from below. Coastal upwelling takes place very strongly in the eastern boundary currents as well as the western coastal areas, which are located along the North American and South African plains.[12] The waters in these areas are rich in nutritional value, approximately two hundred metres of which lies at the surface. This nutritional water lying at the surface stimulates the primary production of nutrients, including organic compounds, and this again results in the deficiency of oxygen in the water below the surface because of these exports of minerals and organic compounds.[13] Moreover, this recycling of organic materials at too much depth results in the accumulation of nutritional elements, leading to nitrate concentrations of 20 to 50 milli-molar.

At higher latitudes, the presence of convection in deep waters causes the mixing of the waters to a certain depth, which again may vary by hundreds of metres. This is the reason why the destruction of seasonal interfaces occurs.[14] So, basically the nutritional elements from the deep waters undergo distribution towards the upper layers, especially the surface layer which provides a reservoir of nutritional elements in the spring bloom of the coming year.[15] Inside the ocean, the formation of certain types of currents or fields is seen, which are known as eddy fields. These can create instabilities, which can be either baro-clinic or baro-tropic. Finally, the formation of gyres may take place, which are large-scale and which are diverse in the horizontal wind field.[16]

2.2 Nitrogen Fixation

Organisms that fix nitrogen are independent of different nitrogenous sources lying in combination with nitrogen dioxide gas, which possess a concentration of more than 400 mmol specifically in the marine world. For the purpose of using nitrogen as dinitrogen, the reduction of nitrogen into ammonium should be carried out.[17] For this conversion, some enzymes are required that possess some specific functions. The organisms that fix nitrogen depend on phosphate and need iron in huge quantities compared to the other phytoplankton.[18] As the input of iron from the land is much lower, especially in the Pacific Ocean when compared to the Atlantic Ocean, the Atlantic Ocean might experience a greater nitrogen fixation than that of the previously regarded Pacific Ocean.[19] The majority of the nitrogen fixation in the marine world is carried out by the genus *Trichodesmium*, which is responsible for the formation of large blooms.[20] The majority of research was focused on this genus only, until recently, when it has shifted towards the small unicellular group of organisms that also possess the ability of nitrogen fixation. These organisms are heterotrophic in nature and their main role is fixing nitrogen.[21] They can be photoheterotroph group-A or UCYN-A, which do not possess photosystem II. They also include *Richelia intracellularis*, which is a symbiotic organism that undergoes nitrogen fixation in diatom cells and is seen specifically in tropical river plumes.[22]

2.3 Stoichiometry of Carbon : Nitrogen : Phosphorus in the Marine Environment

A very strong similarity exists among the ratios of nitrogen and phosphorus content in plankton cells and in the nutritional concentrations of deep-water bodies. Scientists have suggested that an adjustment in the marine life occurs based on its nutritional requirements.[23] The ratio between carbon, nitrogen and phosphorus in the marine environment varies between

106 : 16 : 1 on a molar basis.[24] This ratio is the basic ratio of carbon, nitrogen and phosphorus that exists in every aspect of marine life and is also the basic requirement for the existence of marine organisms and their metabolism.[25] This ratio again links the various elemental cycles, and has been mostly used in the element flux models as well as the ecosystem.[26] Again, the maintenance of this cellular ratio is very important for regulating the organic carbon and nitrogen cycles and linking the activities of single cells and functions specific to that ecosystem. A lower concentration of nitrates is seen in heterotrophic organisms. It is also very important to concentrate on the management of nutritional elements for the maintenance of elemental ratios that are highly balanced, because excess levels of nitrogen ultimately increase the level of nitrates.[27]

2.4 Biogeochemistry of Oxygen Deficiency Zones in the Marine Environment

The oxygen controlling processes of the nitrogen cycle depend on many factors, which include reactions in microorganisms that require oxygen and the inhibition of others by it. The microbiological pathways that ultimately direct the release of fixed nitrogen, denitrification, *etc.*, can take place in close proximity to oxygen or where it is not present at all.[28] These parameters are usually seen in the coastal areas and the shelf sediments across the world in various water columns. They have a comparatively higher utilization rate of oxygen and a much lower ventilation level of oxygen at the same time, and vary by hundreds of metres in thickness.[29] The specific areas in which the concentration of oxygen is very low induce an anaerobic metabolism that is known as the ODZ or oxygen deficient zone of oxygen minimum zones. They experience an oxygen concentration that is usually lower than 8 mmol.[30]

The consistency of the nitrogen cycle in the oxygen deficient zone is mainly a bit complex compared to the consistency in the open and vast ocean. This is because the oxygen deficient zone could be the reason for a huge loss of fixed nitrogen that occurs in the water column.[31] Anammox and denitrification, both of which give rise to the formation of dinitrogen gas, can also lead to the formation of nitric oxide gas. The consumption of DIN is facilitated by both anammox and denitrification, whereas their roles in the carbon cycle are quite different.[32] The bacteria undergoing denitrification usually are heterotrophic bacteria and they take in DIN at an approximate concentration of 70–72 mmol of oxygen. Usually throughout this process, they are responsible for the oxidation of organic compounds. A few denitrifying agents are, however, autotrophic bacterial species and almost all of them fix carbon dioxide and are also responsible for the oxidation of reduced sulphur-based compounds along with nitrate.[33] On the other hand, anammox bacterial species are also autotrophic bacteria that fix carbon dioxide and are responsible for the oxidation of both ammonium and nitrite to dinitrogen and are also responsible for the oxidation of nitrite to nitrate.[34]

The entire process takes place anaerobically. The limitations of both of these processes are that, in oxygen deficient zones of the ocean, both the anammox and the denitrification depend on the constant supply of organic compounds. Without the flow of these, the processes won't occur. Anammox is highly dependent on denitrification for the continuous sourcing of DIN.[35]

2.5 Achievements Regarding Marine Nitrogen

The biogeochemistry of the ocean and the ecosystem models of the ocean possess nitrates as their mode of currency. In the ocean, nitrogen is seen as a state variable. The models that are being used possess substantial levels of variation. In the field of complexity, the formulation of the model as well as the spatial coverage is a critical parameter.[36] Process models, such as models for nitrogen fixation and denitrification models, are created for the investigation of single processes. Circular models of ocean biogeochemistry are designed on the basis of nitrogen phytoplankton *zooplankton detritus* models.[37] They are regarded as the very first models for the consideration of bacteria and DON as state variables. The said models were observed in combination with a circulation model to see an estimation of the basin-wide occurrence of nitrate along with chlorophyll-A and the margins of the various new nitrogen sources.[38] After that, more and more combinations of models were observed and hence they became more and more complex. This again increased the multiple nutritional elements or the functional groups in the plankton involved.[39] The improvement in the production of nitrogen in more and more amounts or production of more and more phosphorus was a revolutionary concept as it gives us a chance to use a combination of the nitrogen, phosphorus and carbon cycles (Table 2.1).[40]

2.6 Impact of Climate Change on Marine Life

A lot of research work conducted earlier has suggested that the change in climate that takes place because of changes in the ocean, the change in the structure of their nitrogen cycles and several other compositions and parameters that exist in the nitrogen cycle, occurs due to an increase in carbon dioxide levels in the atmosphere.[76–79] There are various reasons why this happens. Global warming and acidification are the direct reasons for this. Moreover, the drop in the levels of oxygen in the atmosphere constitutes an indirect reason. This again constitutes a reason that can change the biogeochemistry of the ocean completely.[80–83] Also, an increase in air temperature, leading to warming up of the sea water, causes enhanced activities of ocean stratification as well as microorganisms.[84,85] But neither the activities of microorganisms nor the ocean stratification increase by the same margins in every part of the ocean.[86] In the below or lower latitudes, that is in the Arctic lying between 208N and 208S, there is an occurrence of strong stratification in the water column. However, the other parts of the ocean do not possess this strong level of stratification.[87–89] All of these factors hamper

Table 2.1 Representative studies using N cycling genes and rate measurements of N processes in marine ecosystems: ocean minimum zones, open ocean, deep-sea, and estuarine sediments.

Sl no.	Genes	Rates of N processes	Methods	Major relationships	Conditions tested	Ecosystem	Ref.
1.	AOA-*amoA*, *nxrB*, *nirS*, *Scalindua-nirS*, 16S rRNA	NO_2^- oxidation, NO_3^- and NO_2^- reduction, anammox	Denitrifier and anammox populations mediate low, but significant N loss. Low O_2 supports NO_2^- oxidation, restricting NO_2^- available for anammox or denitrification	HC, qPCR, ^{15}N-IPE	Bay of Bengal OMZ	Evidence of N-loss processes	41
2.	*narG*, 16S rRNA	NO_3^- and NO_2^- reduction	Great abundances of SAR11 lineages at anoxic depths, responsible for ~40% of OMZ *nar* transcripts. SAR11 as important players of NO_2^- production in OMZs	HC, SAG, qPCR, MG, MT	ETNP OMZ	Contribution of SAR11 to N cycling in OMZ	42
3.	Nitrogen fixation, *nifH gene*	NH_3 denitrification and NO_2^- oxidation, anammox	Niche partitioning of N players based on tolerance to O_2. Enrichment of denitrification genes associated with N_2O and N_2 production in particles	HC, ^{15}N-IPE, MG, qPCR, lipid profiles	ETNP OMZ	Relationship of N players between water column and particles at different depths	43

Table 2.1 (*Continued*)

Sl no.	Genes	Rates of N processes	Methods	Major relationships	Conditions tested	Ecosystem	Ref.
4.	Nitrogen fixation, *nifH gene*	NH_3 and NO_2^- oxidation, NO_3^- reduction	HC, ^{15}N-IPE, MG, MT	NO_2^- accumulation in the hypoxic zone due to a decoupling of NH_3 oxidation and NO_2^- oxidation driven by temperature, O_2 and substrate availability	Links among NO_2^- accumulation, microbial taxa and metabolisms across O_2 gradients	Louisiana shelf (Gulf of Mexico)	44
5.	Nitrogen fixation, *nifH gene*	BNF	HC, ^{15}N-incorporation, qPCR, RT-qPCR, pyrosequencing	High diversity of active heterotrophic diazotrophs in anoxic NH_4^+-rich waters	Diversity and activity of heterotrophic diazotrophs	Baltic sea OMZ	45
6.	Major N genes	NH_3 and NO_2^- oxidation, DNRA, denitrification, anammox	HC, ^{15}N-IPE, qPCR	N cycling is linked to the export of organic matter. N-loss is most active over the shelf, fueled by sinking organic matter and benthic NH_4^+ release	Role of organic matter export in N-loss	ETSP OMZ	46
7.	Planctomycetes 16S rRNA	Anammox and denitrification	HC, ^{15}N-IPE, clone libraries, T-RFLP, qPCR	Salinity influences the diversity and abundance of anammox bacteria, which are linked to anammox rates	Spatial-temporal variation in the activity and diversity of anammox bacteria	Cape fear river estuarine sediments	47

8.	*nifH*	BNF	HC, qPCR, clone libraries	Diverse diazotrophs; high and temporal variable BNF in oxic and OMZ denitrified waters	Occurrence of N_2 fixation in denitrified and oxygen-deficient waters	ETSP OMZ	48
9.	*nrfA*, *nirS*, *Scalindua-nirS*	DNRA, anammox, denitrification,	HC, ^{15}N-IPE, RT-qPCR, CARD-FISH, clone libraries	Highly active anammox and hardly detectable denitrification. Anammox coupled to DNRA. N-loss rates linked to organic matter	Mechanisms of biological N_2 production	Arabian OMZ	49
10.	*nirS*, Planctomycetes 16S rRNA	Denitrification, anammox,	HC, ^{15}N-IPE, qPCR	N_2 is mainly produced by denitrification in the Arabian OMZ and by anammox in the ESTP	Activity and abundance of denitrifiers and anammox bacteria	ETSP, Arabian OMZ	50
11.	*amoA*, Crenarchaeota 16S rRNA	NH_3 oxidation	HC, ^{15}N-IPE, qPCR, clone libraries	AOA are the most abundant within OMZ, where nitrification may be coupled to denitrification. AOB are low or undetectable. NH_3 oxidation correlated with AOA	Activity and abundance of NH_3 oxidizers	Gulf of California OMZ	51
12.	Key N genes, 16S rRNA	NH_3 and NO_2^- oxidation, denitrification, anammox	HC, ^{15}N-IPE, 16S Illumina, MG, MT	N cycle activity, except denitrification, is confined to free-living communities that are dependent on particle access	Differences of N metabolic activities between particle-associated and free-living communities	ETNP OMZ	52

Table 2.1 *(Continued)*

Sl no.	Genes	Rates of N processes	Methods	Major relationships	Conditions tested	Ecosystem	Ref.
13.	Planctomycetes 16S rRNA	Anammox	HC, ^{15}N-IPE, FISH, clone libraries, lipid profiles	Anammox is mainly responsible for N_2 loss	Detection of anammox bacteria	Namibian OMZ	53
14.	*nifH*	BNF	HC, ^{15}N-IPE, qPCR, Illumina	Distinct biogeographical patterns among the three regions. Diazotrophs are omnipresent in marine waters, but BNF is regionally restricted	Link between marine diazotrophic diversity and BNF rates	NPSG, CCS, ESP	54
15.	Nitrospina 16S rRNA, *amoA*	NH_3 and NO_2^- oxidation, NO_3^- and NO_2^- reduction	HC, ^{15}N-IPE, IP, qPCR	50% of reduced NO_3^- to NO_2^- is re-oxidized back to NO_3^-. Anammox contributes to NO_2^- oxidation, but is not responsible for all of it	Cycling of NO_3^- and NO_2^- in the secondary NO_2^- maximum	ETNP OMZ	55
16.	*nifH*	BNF	HC, ^{15}N-IPE, MT, MT, qPCR	Different diazotrophic composition between (sub)tropical gyre and cold northern regions, where BNF was also detected	Diazotroph community structure and activity along a 7500 km south-north transect	North Pacific	56

17.	*amoA*	Nitrification	Light experiments, HC, ^{15}N-IPE, qPCR, RT-qPCR	Nitrification in the photic zone is more regulated by competition with phytoplankton for NH_4^+ than by light	Effect of light and phytoplankton growth on the activity of NH_3 oxidizers	Offshore Monterey Bay	57
18.	*amoA*	NH_3 oxidation	HC, ^{15}N-IPE, qPCR, microarrays	Maximum NH_3 oxidation rates at the PNM, where AOA outnumber AOB. Below the PNM AOB outnumber AOA. A diverse AOA community is linked to seasonal biogeochemical changes	Depth distribution of NH_3 oxidation rates and NH_3 oxidizers	Sargasso Sea	58
19.	*amoA*, 16S rRNA of MGI and *Nitrospina*	Nitrification	HC, IP, ^{15}N-IPE, RT-qPCR, clone libraries	AOA outnumber AOB and are distributed in shallow/deep ecotypes. AOA and MGI abundances correlated with *Nitrospina* abundance. Nitrification is not correlated to *amoA* and produces 1.5–4 times the N_2O flux from deep water	Activity and distribution of nitrifiers and their contributions to N cycling	CCS	59

Table 2.1 (*Continued*)

Sl no.	Genes	Rates of N processes	Methods	Major relationships	Conditions tested	Ecosystem	Ref.
20.	Planctomycetes 16S rRNA	Anammox	HC, ^{15}N-IPE, clone libraries, lipid profiles	Active anammox bacteria in hydrothermal vents. Sequences suggest a new anammox clade	Role of anammox in hydrothermal vents	Deep-sea hydrothermal vents	60
21.	*amoA*	Nitrification	Potential oxidation of NH_3, competitive PCR, clone libraries	AOA outnumber AOB. Unique and psychrophilic AOA and AOB are responsible for nitrification in deep cold seep sediments	Diversity and activity of NH_3 oxidizers	Deep-sea cold seep sediments and water	61
22.	*nifH*	BNF, NH_4^+ assimilation	HC, ^{15}N-incorporation, ^{15}N-IPE, clone libraries	High and heterogeneous BNF that is inhibited by NH_4^+. Methane-coupled sulfate reduction dependency of BNF in seep sediments	Activity and distribution of benthic diazotrophs	Deep-sea sediments	62
23.	Planctomycetes 16S rRNA	Anammox	HC, ^{15}N-IPE, FISH, clone libraries	Ubiquitous presence of active anammox bacteria belonging to *Ca.* Scalindua in different anoxic environments	Diversity and activity of anammox bacteria in anoxic marine ecosystems	Anoxic marine sediments and water	63

24.	NC10 16S rRNA, *pmoA*	n-DAMO	HC, IPE, clone libraries, qPCR	Widespread occurrence and high diversity of n-DAMO bacteria, which are more active and abundant in the upper layer	Diversity and activity of n-DAMO bacteria in mangrove sediments	Zhangjiang estuarine sediments	64
25.	*nifH*	BNF	HC, ^{15}N-incorporation, qPCR, RT-qPCR, Illumina	Significant BNF mediated by heterotrophic, photo-heterotrophic and photosynthetic diazotrophs that vary in space and time	Diversity and activity of diazotrophs in two mesohaline temperate estuaries	Roskilde Fjord and the Great Belt strait	65
26.	*napA, narG, nrfA, nirS,* 16S rRNA	Denitrification, DNRA	HC, ^{15}N-IPE, 16S T-RFLP, qPCR, RT-qPCR,	Denitrification, DNRA and corresponding gene abundances and transcripts decrease along the estuary without a temporal correlation	Spatial-temporal variation in the activity and abundance of the denitrifying and DNRA community along a NO_3^- gradient	Colne estuarine sediments	66
27.	*narG, napA, nirS, nrfA*, Planctomycetes 16S rRNA	Denitrification, DNRA, anammox	^{15}N-IPE, NO_3^- reduction potential, qPCR, RT-qPCR, clone libraries	Denitrification and DNRA rates decrease toward the mouth. DNRA potential and *nrfA* increase and	Spatial variation and relationship between gene abundances and rates of denitrification,	Colne estuarine sediments	67

Table 2.1 (*Continued*)

Sl no.	Genes	Rates of N processes	Methods	Major relationships	Conditions tested	Ecosystem	Ref.
				denitrification potential and *nirS* decrease as NO_3^- declines. Anammox only detected at the head	DNRA and anammox		
28.	*amoA,* 16S rRNA	Nitrification	HC, nitrification potential, qPCR, DGGE, 16S pyrosequencing	Nitrification potential differs in space and time. AOB dominate over AOA, with AOB/AOA abundance increasing from the head to the mouth	Spatial-temporal variation in the activity and diversity of AOA and AOB in surface sediments	Colne estuarine sediments	68
29.	*nrfA*	DNRA	HC, ^{15}N-IPE, qPCR, pyrosequencing	Variation of DNRA community along the salinity gradient. DNRA bacterial abundance and organic carbon availability regulate DNRA activity	Activity and diversity of DNRA community along a salinity gradient	New River estuarine sediments	69

30.	Planctomycetes 16S rRNA	Anammox and denitrification	HC, ^{15}N-IPE, clone libraries, qPCR	Salinity defines anammox bacterial distribution, whose activity is related to temperature, nitrite, and anammox bacterial abundance. Anammox is linked to denitrification, which is the main N-loss pathway	Anammox bacterial diversity and activity in marsh sediments	Yangtze estuarine sediments	70
31.	*nirK, nirS*	Denitrification	HC, denitrification potential, qPCR, clone libraries	*nirS* abundance correlated with denitrification and higher than *nirK* abundance at every site and time point. Salinity, organic carbon, N and several metals influence denitrification rates, *nir* abundance and community structure	Spatial-temporal variation in the activity and community structure of denitrifiers	San Francisco estuarine sediments	71

Table 2.1 (*Continued*)

Sl no.	Genes	Rates of N processes	Methods	Major relationships	Conditions tested	Ecosystem	Ref.
32.	*amoA*, *napA*, *narG*, *nirS*, *Scalindua-nirS*, *nrfA*	NH_3 oxidation, DNRA, NO_3^- reduction, anammox	HC, ^{15}N-IPE, qPCR, RT-qPCR	Anammox is the main N-loss pathway and obtains NO_2^- from NO_3^- reduction and aerobic NH_3 oxidation, and NH_4^+ from DNRA and remineralization *via* NO_3^- reduction. Deep-sea NO_3^- accounted for 50% of N loss	Microbial processes responsible for the generation of NH_4^+ and NO_2^- for anammox	ETSP OMZ	72–75

the nutritional supplements when light is present. Also, the reduction in the gas solubility of the ocean, especially for oxygen, leads to a reduction in the structural aspects of oxygen minimum zones.[90]

Ocean acidification might lead to the stimulation of certain species of cyanobacteria that fix nitrogen. In specific regions, in which an increase in the primary production is seen, more and more oxygen is seen to be consumed as the depth increases in the process of remineralization. This causes the release of more and more N_2O into the atmosphere, also causing masses of water to reach the surface of the water. But the rise in temperature might be the reason for the imbalance that may occur between respiration and production resulting in the lower export of organic matter that might take place in the process of counteracting the levels of oxygen loss in the deep ocean.

2.7 Conclusion

Throughout this review, we can see an overall ecological view, genetic view and distribution view of the marine nitrogen cycle and the different forms of nitrogen that are seen in the different types of ocean water. The process of nitrogenous nutrient distribution among the ocean waters is much more complicated than suggested in the earlier research and review reports. This is mainly because of the conclusions made previously regarding marine environments being completely based on speculations. Moreover, with more and more research work being conducted on marine waters and marine life, much more novel nitrogen-involved pathways have been discovered over the last few years. These pathways changed the outline of the marine nitrogen cycle completely. Moreover, with advancements in engineering technologies, human activities that affect marine life and marine waters have increased significantly. This has also hampered the nitrogenous pathways that were functional in the marine environment previously. There are many models and algorithms that can be used to view or to analyse the results of the effect of anthropogenic activities on the marine life and marine nitrogen cycle. But these algorithms and mathematical models are not always one hundred percent correct or accurate because their development has been done based on the previous and present training data. These training data consist of information about the marine nitrogenous pathways and other processes that were present in the past and which are present currently. However, the training data doesn't consist of the probable nitrogenous pathways and processes that will be in the future marine environment. This is the reason why there should be approaches to make different types of holistic mathematical models with the help of the information that is available to us and taking specific parameters into consideration that can change in the future marine environment and marine life.

References

1. G. Abell, A. T. Revill, C. Smith, A. P. Bissett, J. K. Volkman and S. S. Robert, Archaeal ammonia oxidizers and *nirS*-type denitrifiers

dominate sediment nitrifying and denitrifying populations in a subtropical macrotidal estuary, *ISME J.*, 2010, **4**, 286–300.
2. G. Abell, D. Ross, J. Keane, J. Oakes, B. Eyre and S. Robert, *et al.*, Nitrifying and denitrifying microbial communities and their relationship to nutrient fluxes and sediment geochemistry in the Derwent Estuary, Tasmania, *Aquat. Microb. Ecol.*, 2013, **70**, 63–75.
3. N. S. R. Agawin, M. Benavides, A. Busquets, P. Ferriol, L. J. Stal and J. Arístegui, Dominance of unicellular cyanobacteria in the diazotrophic community in the Atlantic Ocean, *Limnol. Oceanogr.*, 2014, **59**, 623–637.
4. H. Agogué, M. Brink, J. Dinasquet and G. J. Herndl, Major gradients in putatively nitrifying and non-nitrifying archaea in the deep North Atlantic, *Nature*, 2008, **456**, 788–791.
5. A. E. Allen, M. G. Booth, M. E. Frischer, P. G. Verity, J. P. Zehr and S. Zani, Diversity and detection of nitrate assimilation genes in marine bacteria, *Appl. Environ. Microbiol.*, 2001, **67**, 5343–5348.
6. L. Alonso-Sáez, A. S. Waller, D. R. Mende, K. Bakker, H. Farnelid and P. L. Yager, *et al.*, Role for urea in nitrification by polar marine Archaea, *Proc. Natl. Acad. Sci. U. S. A.*, 2012, **109**, 17989–17994.
7. D. L. Arévalo-Martínez, A. Kock, C. R. Löscher, R. A. Schmitz and H. W. Bange, Massive nitrous oxide emissions from the Tropical South Pacific Ocean, *Nat. Geosci.*, 2015, **8**, 530–533.
8. D. J. Arp, P. S. G. Chain and M. G. Klotz, The impact of genome analyses on our understanding of ammonia-oxidizing bacteria, *Annu. Rev. Microbiol.*, 2007, **61**, 503–528.
9. S. J. Atkinson, C. G. Mowat, G. A. Reid and S. K. Chapman, An octaheme c-type cytochrome from *Shewanella oneidensis* can reduce nitrite and hydroxylamine, *FEBS Lett.*, 2007, **581**, 3805–3808; A. R. Babbin, D. Bianchi, A. Jayakumar and B. B. Ward, Rapid nitrous oxide cycling in the suboxic ocean, *Science*, 2015, **348**, 1127–1129.
10. A. R. Babbin, R. G. Keil, A. H. Devol and B. B. Ward, Organic matter stoichiometry, flux, and oxygen control nitrogen loss in the ocean, *Science*, 2014, **344**, 406–408.
11. A. R. Babbin, B. D. Peters, C. W. Mordy, B. Widner, K. L. Casciotti and B. B. Ward, Multiple metabolisms constrain the anaerobic nitrite budget in the Eastern Tropical South Pacific, *Global Biogeochem. Cycles*, 2017, **31**, 258–271.
12. S. E. Baer, T. L. Connelly, R. E. Sipler, P. L. Yager and D. A. Bronk, Effect of temperature on rates of ammonium uptake and nitrification in the western coastal Arctic during winter, spring, and summer, *Global Biogeochem. Cycles*, 2014, **28**, 1455–1466.
13. B. J. Baker, R. A. Lesniewski and G. J. Dick, Genome-enabled transcriptomics reveals archaeal populations that drive nitrification in a deep-sea hydrothermal plume, *ISME J.*, 2012, **6**, 2269–2279.
14. B. J. Baker, C. S. Sheik, C. A. Taylor, S. Jain, A. Bhasi and J. D. Cavalcoli, *et al.*, Community transcriptomic assembly reveals microbes that contribute to deepsea carbon and nitrogen cycling, *ISME J.*, 2013, **7**, 1962–1973.

15. N. J. Bale, L. Villanueva, H. Fan, L. J. Stal, E. C. Hopmans and S. Schouten, *et al.*, Occurrence and activity of anammox bacteria in surface sediments of the southern North Sea, *FEMS Microbiol. Ecol.*, 2014, **89**, 99–110.
16. M. Bandekar, N. Ramaiah and R. M. Meena, Diversity and abundance of denitrifying and anammox bacteria from the Arabian Sea oxygen minimum zone, *Deep Sea Res., Part II*, 2018, **156**, 19–26.
17. H. W. Bange, A. Freing, A. Kock and C. R. Löscher, Marine pathways to nitrous oxide, in *Nitrous Oxide and Climate Change*, ed. K. Smith, Earthscan, London, 2010, pp. 36–62.
18. G. Battaglia and F. Joos, Marine N_2O emissions from nitrification and denitrification constrained by modern observations and projected in multimillennial global warming simulations, *Global Biogeochem. Cycles*, 2018, **32**, 92–121.
19. J. M. Beman and C. A. Francis, Diversity of ammonia-oxidizing archaea and bacteria in the sediments of a hypernutrified subtropical estuary: bahia del Tobari, Mexico, *Appl. Environ. Microbiol.*, 2006, **72**, 7767–7777.
20. J. M. Beman, C.-E. Chow, A. L. King, Y. Feng, J. A. Fuhrman and A. Andersson, *et al.*, Global declines in oceanic nitrification rates as a consequence of ocean acidification, *Proc. Natl. Acad. Sci. U. S. A.*, 2011, **108**, 208–213.
21. J. M. Beman, J. Leilei Shih and B. N. Popp, Nitrite oxidation in the upper water column and oxygen minimum zone of the Eastern Tropical North Pacific Ocean, *ISME J.*, 2013, **7**, 2192–2205.
22. J. M. Beman, B. N. Popp and S. E. Alford, Quantification of ammonia oxidation rates and ammonia-oxidizing archaea and bacteria at high resolution in the Gulf of California and Eastern Tropical North Pacific Ocean, *Limnol. Oceanogr.*, 2012, **57**, 711–726.
23. J. M. Beman, B. N. Popp and C. A. Francis, Molecular and biogeochemical evidence for ammonia oxidation by marine crenarchaeota in the Gulf of California, *ISME J.*, 2008, **2**, 429–441.
24. J. M. Beman, K. J. Roberts, L. Wegley, F. Rohwer and C. A. Francis, Distribution and diversity of archaeal ammonia monooxygenase genes associated with corals, *Appl. Environ. Microbiol.*, 2007, **73**, 5642–5647.
25. J. M. Beman, R. Sachdeva and J. A. Fuhrman, Population ecology of nitrifying archaea and bacteria in the Southern California bight, *Environ. Microbiol.*, 2010, **12**, 1282–1292.
26. M. Benavides, V. N. Bednarz and C. Ferrier-Pagès, Diazotrophs: overlooked key players within the coral symbiosis and tropical reef ecosystems?, *Front. Mar. Sci.*, 2017, **4**, 10.
27. M. Benavides, H. Berthelot, S. Duhamel, P. Raimbault and S. Bonnet, Dissolved organic matter uptake by *Trichodesmium* in the Southwest Pacific, *Sci. Rep.*, 2017, **7**, 41315.
28. M. Benavides, S. Bonnet, N. Hernández, A. M. Martínez-Pérez, M. Nieto-Cid and X. A. Álvarez-Salgado, *et al.*, Basin-wide N_2 fixation in the deep

waters of the Mediterranean Sea, *Global Biogeochem. Cycles*, 2016, **30**, 952–961.
29. M. Benavides, F. Houlbrèque, M. Camps, A. Lorrain, O. Grosso and S. Bonnet, Diazotrophs: a non-negligible source of nitrogen for the tropical coral Stylophora pistillata, *J. Exp. Biol.*, 2016, **219**, 2608–2612.
30. M. Benavides, P. H. Moisander, H. Berthelot, T. Dittmar, O. Grosso and S. Bonnet, Mesopelagic N_2 fixation related to organic matter composition in the Solomon and Bismarck Seas (Southwest Pacific), *PLoS One*, 2015, **10**, e0143775.
31. M. Bentzon-Tilia, S. J. Traving, M. Mantikci, H. Knudsen-Leerbeck, J. L. S. Hansen and S. Markager, *et al.*, Significant N_2 fixation by heterotrophs, photoheterotrophs and heterocystous cyanobacteria in two temperate estuaries, *ISME J.*, 2015, **9**, 273–285.
32. C. Berg, V. Vandieken, B. Thamdrup and K. Jürgens, Significance of archaeal nitrification in hypoxic waters of the Baltic Sea, *ISME J.*, 2015, **9**, 1319–1332.
33. B. Bergman and E. J. Carpenter, Nitrogenase confined to randomly distributed trichomes in the marine cyanobacterium *Trichodesmium thiebautii*, *J. Phycol.*, 1991, **27**, 158–165.
34. I. Berman-Frank, J. T. Cullen, Y. Shaked, R. M. Sherrell and P. G. Falkowski, Iron availability, cellular iron quotas, and nitrogen fixation in *Trichodesmium*, *Limnol. Oceanogr.*, 2001, **46**, 1249–1260.
35. I. Berman-Frank, P. Lundgren, Y.-B. Chen, H. Küpper, Z. Kolber and B. Bergman, *et al.*, Segregation of nitrogen fixation and oxygenic photosynthesis in the marine cyanobacterium *Trichodesmium*, *Science*, 2001, **294**, 1534–1537.
36. A. E. Bernhard, T. Donn, A. E. Giblin and D. A. Stahl, Loss of diversity of ammonia-oxidizing bacteria correlates with increasing salinity in an estuary system, *Environ. Microbiol.*, 2005, **7**, 1289–1297.
37. A. E. Bernhard, Z. C. Landry, A. Blevins, J. R. de la Torre, A. E. Giblin and D. A. Stahl, Abundance of ammonia-oxidizing archaea and bacteria along an estuarine salinity gradient in relation to potential nitrification rates, *Appl. Environ. Microbiol.*, 2010, **76**, 1285–1289.
38. A. D. Bertagnolli and O. Ulloa, Hydrography shapes community composition and diversity of amoA-containing thaumarchaeota in the coastal waters off central Chile, *Environ. Microbiol. Rep.*, 2017, **9**, 717–728.
39. H. Berthelot, M. Benavides, P. H. Moisander, O. Grosso and S. Bonnet, High-nitrogen fixation rates in the particulate and dissolved pools in the Western Tropical Pacific (Solomon and Bismarck Seas), *Geophys. Res. Lett.*, 2017, **44**, 8414–8423.
40. M. Blais, J.-E. Tremblay, A. D. Jungblut, J. Gagnon, J. Martin and M. Thaler, *et al.*, Nitrogen fixation and identification of potential diazotrophs in the Canadian Arctic, *Global Biogeochem. Cycles*, 2012, **26**, GB3022.
41. D. Bombar, R. W. Paerl and L. Riemann, Marine non-cyanobacterial diazotrophs: moving beyond molecular detection, *Trends Microbiol.*, 2016, **24**, 916–927.

42. S. Bonnet, I. C. Biegala, P. Dutrieux, L. O. Slemons and D. G. Capone, Nitrogen fixation in the Eestern Equatorial Pacific: rates, diazotrophic cyanobacterial size class distribution, and biogeochemical significance, *Global Biogeochem. Cycles*, 2009, **23**, GB3012.
43. S. Bonnet, J. Dekaezemacker, K. A. Turk-Kubo, T. Moutin, R. M. Hamersley and O. Grosso, *et al.*, Aphotic N_2 fixation in the Eastern Tropical South Pacific Ocean, *PLoS One*, 2013, **8**, e81265.
44. S. Bonnet, M. Rodier, K. A. Turk-Kubo, C. Germineaud, C. Menkes and A. Ganachaud, *et al.*, Contrasted geographical distribution of N_2 fixation rates and *nifH* phylotypes in the Coral and Solomon Seas (southwestern Pacific) during austral winter conditions, *Global Biogeochem. Cycles*, 2015, **29**, 1874–1892.
45. D. Böttjer, D. M. Karl, R. M. Letelier, D. A. Viviani and M. J. Church, Experimental assessment of diazotroph responses to elevated seawater pCO_2 in the North Pacific subtropical gyre, *Global Biogeochem. Cycles*, 2014, **28**, 601–616.
46. A. Bourbonnais, S. K. Juniper, D. A. Butterfield, R. E. Anderson and M. F. Lehmann, Diversity and abundance of bacteria and *nirS*-encoding denitrifiers associated with the juan de fuca ridge hydrothermal system, *Ann. Microbiol.*, 2014, **64**, 1691–1705.
47. A. Bourbonnais, R. T. Letscher, H. W. Bange, V. Échevin, J. Larkum and J. Mohn, *et al.*, N_2O production and consumption from stable isotopic and concentration data in the peruvian coastal upwelling system, *Global Biogeochem. Cycles*, 2017, **31**, 678–698.
48. N. J. Bouskill, D. Eveillard, D. Chien, A. Jayakumar and B. B. Ward, Environmental factors determining ammonia-oxidizing organism distribution and diversity in marine environments, *Environ. Microbiol.*, 2012, **14**, 714–729.
49. M. W. Bowles, L. M. Nigro, A. P. Teske and S. B. Joye, Denitrification and environmental factors influencing nitrate removal in Guaymas Basin hydrothermally altered sediments, *Front. Microbiol.*, 2012, **3**, 377.
50. G. Braker and J. M. Tiedje, Nitric oxide reductase (*norB*) genes from pure cultures and environmental samples, *Appl. Environ. Microbiol.*, 2003, **69**, 3476–3483.
51. G. Braker, J. Zhou, L. Wu, A. H. Devol and J. M. Tiedje, Nitrite reductase genes (*nirK* and *nirS*) as functional markers to investigate diversity of denitrifying bacteria in Pacific northwest marine sediment communities, *Appl. Environ. Microbiol.*, 2000, **66**, 2096–2104.
52. J. Brandsma, J. van de Vossenberg, N. Risgaard-Petersen, M. C. Schmid, P. Engström and K. Eurenius, *et al.*, A multi-proxy study of anaerobic ammonium oxidation in marine sediments of the Gullmar Fjord, Sweden, *Environ. Microbiol. Rep.*, 2011, **3**, 360–366.
53. L. D. Brin, A. E. Giblin and J. J. Rich, Environmental controls of anammox and denitrification in southern New England estuarine and shelf sediments, *Limnol. Oceanogr.*, 2014, **59**, 851–860.

54. L. A. Bristow, C. M. Callbeck, M. Larsen, M. A. Altabet, J. Dekaezemacker and M. Forth, *et al.*, N_2 production rates limited by nitrite availability in the Bay of Bengal oxygen minimum zone, *Nat. Geosci.*, 2017, **10**, 24–29.
55. L. A. Bristow, T. Dalsgaard, L. Tiano, D. B. Mills, A. D. Bertagnolli and J. J. Wright, *et al.*, Ammonium and nitrite oxidation at nanomolar oxygen concentrations in oxygen minimum zone waters, *Proc. Natl. Acad. Sci. U. S. A.*, 2016, **113**, 10601–10606.
56. L. A. Bristow, N. Sarode, J. Cartee, A. Caro-Quintero, B. Thamdrup and F. J. Stewart, Biogeochemical and metagenomic analysis of nitrite accumulation in the Gulf of Mexico hypoxic zone, *Limnol. Oceanogr.*, 2015, **60**, 1733–1750.
57. C. Bu, Y. Wang, C. Ge, H. A. Ahmad, B. Gao and S.-Q. Ni, Dissimilatory nitrate reduction to ammonium in the yellow river estuary: rates, abundance, and community diversity, *Sci. Rep.*, 2017, **7**, 6830.
58. C. Buchwald, A. E. Santoro, R. H. R. Stanley and K. L. Casciotti, Nitrogen cycling in the secondary nitrite maximum of the Eastern Tropical North Pacific off Costa Rica, *Global Biogeochem. Cycles*, 2015, **29**, 2061–2081.
59. A. J. Burgin and S. K. Hamilton, Have we overemphasized the role of denitrification in aquatic ecosystems? A review of nitrate removal pathways, *Front. Ecol. Environ.*, 2007, **5**, 89–96.
60. J. A. Burns, J. P. Zehr and D. G. Capone, Nitrogen-fixing phylotypes of chesapeake bay and neuse river estuary sediments, *Microb. Ecol.*, 2002, **44**, 336–343.
61. N. Byrne, M. Strous, V. Crépeau, B. Kartal, J.-L. Birrien and M. Schmid, *et al.*, Presence and activity of anaerobic ammonium-oxidizing bacteria at deep-sea hydrothermal vents, *ISME J.*, 2009, **3**, 117–123.
62. A. M. Cabello, F. M. Cornejo-Castillo, N. Raho, D. Blasco, M. Vidal and S. Audic, *et al.*, Global distribution and vertical patterns of a prymnesiophyte–cyanobacteria obligate symbiosis, *ISME J.*, 2016, **10**, 693–706.
63. J. M. Caffrey, N. Bano, K. Kalanetra and J. T. Hollibaugh, Ammonia oxidation and ammonia-oxidizing bacteria and archaea from estuaries with differing histories of hypoxia, *ISME J.*, 2007, **1**, 660–662.
64. H. Cao, Y. Hong, M. Li and J. D. Gu, Diversity and abundance of ammonia-oxidizing prokaryotes in sediments from the coastal Pearl River Estuary to the South China Sea, *Antonie van Leeuwenhoek*, 2011, **100**, 545–556.
65. H. Cao, Y. Hong, M. Li and J. D. Gu, Lower abundance of ammonia-oxidizing archaea than ammonia-oxidizing bacteria detected in the subsurface sediments of the northern South China Sea, *Geomicrobiol. J.*, 2012, **29**, 332–339.
66. H. Cao, Z. Shao, J. Li, W. Zhang and P. Y. Qian, Phylogenetic diversity of nitrogen-utilizing genes in hydrothermal chimneys from 3 middle ocean ridges, *Extremophiles*, 2015, **19**, 1173–1182.
67. D. G. Capone, *Trichodesmium*, a globally significant marine cyanobacterium, *Science*, 1997, **276**, 1221–1229.

68. D. G. Capone, J. A. Burns, J. P. Montoya, A. Subramaniam, C. Mahaffey and T. Gunderson, *et al.*, Nitrogen fixation by *Trichodesmium* spp.: an important source of new nitrogen to the tropical and subtropical North Atlantic Ocean, *Global Biogeochem. Cycles*, 2005, **19**, GB2024.
69. D. Capone, A. Subramaniam, J. Montoya, M. Voss, C. Humborg and A. Johansen, *et al.*, An extensive bloom of the N_2-fixing cyanobacterium *Trichodesmium erythraeum* in the central Arabian Sea, *Mar. Ecol.: Prog. Ser.*, 1998, **172**, 281–292.
70. J. D. Caranto and K. M. Lancaster, Nitric oxide is an obligate bacterial nitrification intermediate produced by hydroxylamine oxidoreductase, *Proc. Natl. Acad. Sci. U. S. A.*, 2017, **114**, 8217–8222.
71. J. D. Caranto, A. C. Vilbert and K. M. Lancaster, *Nitrosomonas europaea* cytochrome P460 is a direct link between nitrification and nitrous oxide emission, *Proc. Natl. Acad. Sci. U. S. A.*, 2016, **113**, 14704–14709.
72. U. Cardini, V. N. Bednarz, M. S. Naumann, N. van Hoytema, L. Rix and R. A. Foster, *et al.*, Functional significance of dinitrogen fixation in sustaining coral productivity under oligotrophic conditions, *Proc. Biol. Sci.*, 2015, **282**, 20152257.
73. J. Cardoso, J. van Bleijswijk, H. Witte and F. van Duyl, Diversity and abundance of ammonia-oxidizing archaea and bacteria in tropical and coldwater coral reef sponges, *Aquat. Microb. Ecol.*, 2013, **68**, 215–230.
74. P. Carini, C. L. Dupont and A. E. Santoro, Patterns of thaumarchaeal gene expression in culture and diverse marine environments, *Environ. Microbiol.*, 2018, **20**, 2112–2124.
75. K. L. Casciotti and C. Buchwald, Insights on the marine microbial nitrogen cycle from isotopic approaches to nitrification, *Front. Microbiol.*, 2012, **3**, 356.
76. K. L. Casciotti and B. B. Ward, Phylogenetic analysis of nitric oxide reductase gene homologues from aerobic ammonia-oxidizing bacteria, *FEMS Microbiol. Ecol.*, 2005, **52**, 197–205.
77. K. L. Casciotti, M. Forbes, J. Vedamati, B. D. Peters, T. S. Martin and C. W. Mordy, Nitrous oxide cycling in the Eastern Tropical South Pacific as inferred from isotopic and isotopomeric data, *Deep Sea Res., Part II*, 2018, **156**, 155–167.
78. M. Castro-Gonzalez, G. Braker, L. Farias and O. Ulloa, Communities of *nirS*-type denitrifiers in the water column of the oxygen minimum zone in the Eastern South Pacific, *Environ. Microbiol.*, 2005, **7**, 1298–1306.
79. M. Castro-González, V. Molina, E. Rodríguez-Rubio and O. Ulloa, The first report of a microdiverse anammox bacteria community in waters of Colombian Pacific, a transition area between prominent oxygen minimum zones of the Eastern Tropical Pacific, *Environ. Microbiol. Rep.*, 2014, **6**, 595–604.
80. P. Chain, J. Lamerdin, F. Larimer, W. Regala, V. Lao and M. Land, *et al.*, Complete genome sequence of the ammonia-oxidizing bacterium and obligate chemolithoautotroph *Nitrosomonas europaea*, *J. Bacteriol.*, 2003, **185**, 2759–2773.

81. B. X. Chang, J. R. Rich, A. Jayakumar, H. Naik, A. K. Pratihary and R. G. Keil, *et al.*, The effect of organic carbon on fixed nitrogen loss in the eastern tropical South Pacific and Arabian Sea oxygen deficient zones, *Limnol. Oceanogr.*, 2014, **59**, 1267–1274.
82. J. Chen, R. Dick, J.-G. Lin and J.-D. Gu, Current advances in molecular methods for detection of nitrite-dependent anaerobic methane oxidizing bacteria in natural environments, *Appl. Microbiol. Biotechnol.*, 2016, **100**, 9845–9860.
83. J. Chen, X.-W. W. Jiang and J.-D. D. Gu, Existence of novel phylotypes of nitrite-dependent anaerobic methane-oxidizing bacteria in surface and subsurface sediments of the South China Sea, *Geomicrobiol. J.*, 2015, **32**, 1–10.
84. J. Chen, Z. Zhou and J.-D. Gu, Complex community of nitritedependent anaerobic methane oxidation bacteria in coastal sediments of the Mai Po wetland by PCR amplification of both 16S rRNA and *pmoA* genes, *Appl. Microbiol. Biotechnol.*, 2015, **99**, 1463–1473.
85. J. Chen, Z.-C. Zhou and J.-D. Gu, Occurrence and diversity of nitritedependent anaerobic methane oxidation bacteria in the sediments of the South China Sea revealed by amplification of both 16S rRNA and *pmoA* genes, *Appl. Microbiol. Biotechnol.*, 2014, **98**, 5685–5696.
86. S. Cheung, K. Suzuki, H. Saito, Y. Umezawa, X. Xia and H. Liu, Highly heterogeneous diazotroph communities in the kuroshio current and the Tokara Strait Japan, *PLoS One*, 2017, **12**, e0186875.
87. G. D. Christman, M. T. Cottrell, B. N. Popp, E. Gier and D. L. Kirchman, Abundance, diversity, and activity of ammonia-oxidizing prokaryotes in the coastal arctic ocean in summer and winter, *Appl. Environ. Microbiol.*, 2011, **77**, 2026–2034.
88. P. M. Chronopoulou, F. Shelley, W. J. Pritchard, S. T. Maanoja and M. Trimmer, Origin and fate of methane in the Eastern Tropical North Pacific oxygen minimum zone, *ISME J.*, 2017, **11**, 1386–1399.
89. M. J. Church, B. Wai, D. M. Karl and E. F. DeLong, Abundances of crenarchaeal *amoA* genes and transcripts in the Pacific Ocean, *Environ. Microbiol.*, 2010, **12**, 679–688.
90. P. Ciais, C. Sabine, G. Bala, L. Bopp, V. Brovkin and J. Canadell, *et al.*, Carbon and other biogeochemical cycles, in *Climate Change 2013: The Physical Science Basis. Contribution of Working Group I to the Fifth Assessment Report of the Intergovernmental Panel on Climate Change*, ed. T. F. Stocker, D. Qin, G. K. Plattner, M. Tignor, S. K. Allen and J. Boschung, *et al.*, IPCC, Geneva, 2013, pp. 465–570.
91. C. J. Coates and M. Wyman, A denitrifying community associated with a major, marine nitrogen fixer, *Environ. Microbiol.*, 2017, **19**, 4978–4992.
92. L. A. Codispoti, An oceanic fixed nitrogen sink exceeding 400 Tg N a^{-1} *vs.* the concept of homeostasis in the fixed-nitrogen inventory, *Biogeosciences*, 2007, **4**, 233–253.
93. L. A. Codispoti, J. A. Brandes, J. P. Christensen, A. H. Devol, S. W. A. Naqvi and H. W. Paerl, *et al.*, The oceanic fixed nitrogen and

nitrous oxide budgets: moving targets as we enter the anthropocene?, *Sci. Mar.*, 2001, **65**, 85–105.
94. H. Daims, E. V. Lebedeva, P. Pjevac, P. Han, C. Herbold and M. Albertsen, *et al.*, Complete nitrification by Nitrospira bacteria, *Nature*, 2015, **528**, 504–509.
95. H. Daims, S. Lücker and M. Wagner, A new perspective on microbes formerly known as nitrite-oxidizing bacteria, *Trends Microbiol.*, 2016, **24**, 699–712.
96. O. R. Dale, C. R. Tobias and B. Song, Biogeographical distribution of diverse anaerobic ammonium oxidizing (anammox) bacteria in Cape Fear River Estuary, *Environ. Microbiol.*, 2009, **11**, 1194–1207.

CHAPTER 3

Anammox Bacteria and Their Application in Wastewater Treatment Plants

SATARUPA DEY,*[a] ABHIJIT DEY,[b] UTTPAL ANAND[c] AND VINEET KUMAR[d]

[a] Department of Botany, Shyampur Siddheswari Mahavidyalaya, Howrah, West Bengal, 711312, India; [b] Department of Life Sciences, Presidency University, 86/1, College Street, Kolkata, West Bengal, 700073, India; [c] Zuckerberg Institute for Water Research, Jacob Blaustein Institutes for Desert Research, Ben-Gurion University of the Negev, Sede Boqer Campus, Midreshet Ben-Gurion 8499000, Israel; [d] Department of Microbiology, School of Life Sciences, Central University of Rajasthan, Bandarsindri, Ajmer-305817, Rajasthan, India
*Email: dey1919@gmail.com

3.1 Introduction

The removal and treatment of nitrogen-containing waste from wastewater plants is of great importance. Ammonia is the most common nitrogen-containing pollutant present in municipal and industrial wastewater and agrochemicals, and negatively impacts aquatic ecosystems. A higher concentration of ammonia leads to the excessive growth of harmful algae, impacting the quality of water and jeopardising human health.[1] Excessive toxic algal growth and eutrophication reduces the dissolved oxygen and produces cyanotoxins, which are harmful to the aquatic fauna. Moreover, the consumption of fish from polluted water has adverse health effects in humans due to ammonia.[2] Therefore, there are stringent rules and regulations and

Ammonia Oxidizing Bacteria: Applications in Industrial Wastewater Treatment
Edited by Maulin P. Shah

Published by the Royal Society of Chemistry, www.rsc.org

all industries are forced to follow them to eliminate ammonia from wastewater. Although a wide range of physical, chemical, and biological approaches are available to remove ammonia from wastewater, it is necessary that these methods are sustainable and economical so that can be widely accepted. Physiochemical processes such as membrane filtration, which includes processes like reverse osmosis, microfiltration, ultrafiltration, and nanofiltration, and also ion exchange using natural minerals, ozone oxidation and chemical precipitation[1,3] are used for wastewater treatment. The main drawbacks of these processes are high operational cost and skilled labor requirement.[2] Thus, much attention is given to biological approaches, which are cost effective and eco-friendly. Anaerobic ammonium oxidation (Anammox) is a biological approach for nitrogen removal that uses anammox bacteria, involving the process of direct oxidation of ammonium to form dinitrogen gas (N_2). This method is considered as an environmentally friendly and effective technique to treat ammonia-rich wastewater.[4] Anammox bacteria are autotrophic in nature and use ammonium as a source of energy and also directly reduce nitrite into N_2 while producing nitrate as a by-product.[5] This anammox process accounts for nearly 50% of nitrogen recycling in oceanic environments under diverse environmental conditions, including different salinities and temperatures.[6] It has also gained importance because of the higher costs associated with conventional nitrification and denitrification to remove nitrogen, and recently municipal wastewater, pharmaceutical wastewater, and landfill leachates have been biologically treated using the anammox process. However, a detailed understanding of the metabolism, cell biology, ecology and growth physiology of anammox bacteria is needed for their successful application in industry. Moreover, studies have been conducted to gain an idea regarding the impact of specific environments, chemicals, and physiological conditions, for better application of the anammox process in industry for the treatment of ammonia-rich waste streams.[6] This process also has several other advantages, including no external carbon requirement or aeration, which can reduce the carbon footprint of wastewater treatment.[6] The current chapter highlights the characteristics of anammox bacteria and provides key insights into the functional genes involved in the metabolism and oxidation of ammonia. Moreover, the use of anammox bacteria in wastewater treatment is discussed in detail, along with their future prospects and the successful utilization of the anammox process in wastewater treatment.

3.2 The Discovery of Anammox Bacteria

Anaerobic ammonium oxidation (anammox) bacteria were discovered as early as 1990,[5,7] and they show a unique metabolic activity that can combine ammonium and nitrite or nitrate to form nitrogen gas. These anammox bacteria are known to play a crucial role in the enormous loss of nitrogen and are of great importance as anammox can serve as a cost-effective and environment-friendly nitrogen removal process from wastewater. Several studies have been conducted that focus on the application of these anammox bacteria in laboratory scale conditions and reactors, and they have immense application potential on an industrial scale. The discovery of anammox bacteria paved the way towards the development of new biochemical pathways related to inorganic nitrogen removal.

The anammox process is largely done by anammox bacteria and the first anammox bacteria identified was "*Brocadia anammoxidans*". This identification was not done according to the standard microbiological processes and was given a status *Candidatus*, having 16S rRNA gene sequence identities of the species ranging between 87 and 99% and showing large phylogenetic distance. However, all anammox bacteria belong to the same monophyletic cluster Brocadiales and are related to the order Planctomycetales. To date, 6 genera are known, which are affiliated to the phylum Planctomycetes, namely *Candidatus Scalindua*,[8] *Candidatus Brocadia*,[9] *Candidatus Kuenenia*,[8] *Candidatus Anammoxoglobus*,[10] and *Candidatus Jettenia*.[11] They are mostly slow growers and pure cultures are difficult to find. Due to their inability to grow in different available media, they are mostly studied by culture-independent methods from different ecological conditions and anthropogenic impacted environments, and are visible as bright red granules.[12] Oshiki *et al.*[13] have reported a total of 19 species under 6 Candidatus genera (*Brocadia*, *Kuenenia*, *Jettenia*, *Scalindua*, *Anammoxoglobus* and *Anammoximicrobium*) which are predominant both in marine environments and from the sludge of wastewater treatment plants and wastewater-impacted environments.[14] Anammox bacteria are found ubiquitously in anoxic environments in marine situations and suboxic water columns[15] and act as key players in the global nitrogen cycle.

3.2.1 Morphology of Anammox Bacteria

Generally, all bacterial cells can be classified into two classes, Gram-negative and Gram-positive, depending on their cell envelope structure. Anammox bacteria comprise largely of members of Planctomycetes, which have a complex compartmentalization and are largely made up of a single intracytoplasmic membrane.[16] They lack the universal bacterial peptidoglycan layer and have been studied thoroughly by electron microscopy observations, chemical analysis, genome sequencing and resistance to various types of antibiotics. They have two membranes on the inner side and no membrane on the outer side of the cell wall, which is proteinaceous in nature. The organization of the cellular compartment is different from any Gram-negative cells.[17] The cytoplasm of these bacteria is divided into three different cytoplasmic compartments,[18] namely (i) the outer region or paryphoplasm, which is situated between the cytoplasmic membrane and intracytoplasmic membrane, (ii) the riboplasm, which contains DNA, ribosomes and glycogen granules, and (iii) the inner compartment, which is riboplasm-free, known as the anammoxosome and comprises 50–70% of the total cell volume. The membranes of anammox bacteria is composed of glycerolipid bilayers, where the glycerol moieties are linked to fatty acid residues both *via* ester bonds and ether bonds. Extraordinary lipid structures are observed with 3 or 5 linearly concatenated cyclobutane rings, which are known as [3]- and [5]-ladderanes, respectively, which also contain phosphocholine (PC) and phosphoethanolamine (PE). Ladderanes are reported to form a dense membrane around the anammoxosome.

3.2.2 Biofilm Forming Ability

In general, anammox bacteria are known to form aggregates or biofilms having different types of microenvironment due to high microbial activity and nutrient availability.[19] These associations comprise small microbial flocs, granules, and thin biofilms,[20] and this association makes it very difficult to understand the mechanisms or factors involved in niche differentiation. The size of microcolony, density of biomass and spatial distribution are dependent on the half-saturation coefficient in activated sludge flocs. These biofilms also provide a protective environment by forming a physical barrier against washouts, pathogens, oxygen and any toxic chemical.[21] Ali *et al.*[19] have quantitatively evaluated the aggregation abilities of three different anammox species, including *Candidatus Brocadia sinica*, *Candidatus Jettenia caeni* and *Candidatus Brocadia sapporoensis* on the basis of cell surface properties and extracellular polymeric substance (EPS) production. Also, it was reported that *Ca. B. sinica* showed the highest growth rate,[22] which suggests its superiority in wastewater treatment.

3.2.3 Metabolism of Anammox Bacteria

Most anammox cells display a very complex cell architecture having central compartment-like membranes and under a microscope they appear as small coccoid cells with a diameter of approximately 800 nm. They are slow growers having low metabolic activity (forming 15–80 μmol of N_2 per g dry weight of cells per min) and the number of cells is found to double in 11 to 20 days. They are found to thrive on a very low substrate concentration. They are obligate anaerobes and their metabolism is reversibly inhibited by an increase in oxygen concentration.[23] The sequencing batch reactor (SBR) is used to maintain and optimize the enrichment study of anammox bacteria. Stable enrichments are usually established after 90–200 days, after being sparged with oxygen-free gas and inoculated with a suitable wastewater sludge, river or marine sediment.[9,10] The reactors are fed by ammonium, nitrite, bicarbonate and nitrate to avoid low redox potentials. During the 90–200 days of enrichment conditions, the content of the reactor turns red due to the presence of heme proteins, which is associated with a 70% increase in anammox bacteria population. Bicarbonate acts as a carbon source for cell biomass synthesis and carbon fixation is carried out by the acetyl-CoA pathway.[24] It is generally hypothesized that reducing equivalents for the reduction of CO_2 are derived from the oxidation of nitrite to nitrate.[25] Both ammonium and nitrite were reported to be utilized until their concentration is very low. Anammox bacteria are usually not strictly chemolithoautotrophic in nature but are known to have a more versatile lifestyle. They are also reported to use ferrous iron apart from ammonium and also a wide range of organic compounds such as formate, propionate, acetate and methyamines.[9,10] Moreover, they are also found to utilize manganese oxides as electron acceptors in their metabolism.[26] For denitrification, these communities use a different pathway in which nitrate is reduced to nitrite and ammonium and finally to N_2.[10] Kartal *et al.*[9,10] reported that "*Candidatus Anammoxoglobus propionicus*" and "*Candidatus Brocadia fulgida*" occupied different niches and show species-specific adaptive mechanisms under identical conditions in the SBR. "*Candidatus B. fulgida*" specifically became the dominant anammox species when the reactor was amended with acetate, whereas "*Candidatus A. propionicus*" was dominant in the SBR fed with propionate, which provides the cells with a competitive advantage.

3.2.4 Energy Metabolism in Anammox

The immunogold localization of hydrazine/hydroxylamine oxidoreductase HAO has shown that the anammox reaction occurs in the anammoxosome compartment, which results in the formation of proton motive force.[18] However, there is no direct link between this protein and other proteins of the electron transport chain, which actively translocate protons across the anammoxosome membrane. The protons then flow back to the riboplasm through ATPases, which generates ATP. Four different gene clusters were found for ATPases; however, the relationship and location of anammox ATPases is still not very clear. A cytochrome peroxidase staining study detected cytochrome c proteins to be associated with the anammoxosome membrane,[18] showing their involvement in the electron transport chain in the anammoxosome membrane.

3.2.5 Oxygen Tolerance of Anammox Bacteria

Anammox bacteria are known to anoxically oxidize ammonium, where nitrite acts as an electron acceptor, which is supplied by ammonia-oxidizing bacteria or archaea thriving in the same community. This property of anammox bacteria plays a decisive role in their association with ammonia-oxidizing bacteria or archaea. To date, very limited information is available on the oxygen tolerance of anammox bacteria. Straka *et al.*[26] reported that anammox bacteria were inhibited by oxygen at very low levels; however, they are found to have higher KNH_4^+ values than ammonia-oxidizing archaea and almost comparable to that of ammonia-oxidizing bacteria.

3.2.6 Genome Structure of Anammox Bacteria

Strous *et al.*[24] for the first time sequenced the genome of *Kuenenia stuttgartiensis* from which five supercontigs of 4.2 Mb have been identified, which are related to essential genes in major biosynthetic pathways, DNA replication, transcription, translation and protein translocation.

Kuenenia stuttgartiensis has a chemolithoautotrophic mode of nutrition; however, the pathway of carbon fixation cannot be predicted properly. The genome of *K. stuttgartiensis* codes for a complete acetyl-CoA (Wood–Ljungdahl) pathway, while all other known carbon fixation pathways are either missing or incomplete.[24] In *K. stuttgartiensis*, two molecules of CO_2 are reduced and bound to coenzyme A to form acetyl-CoA with the help of two key enzymes, formate dehydrogenase and carbon monoxide dehydrogenase.[24] The one carbon metabolism of *K. stuttgartiensis* involves methanopterin and genes related to methanopterin are found in related Planctomycetes.[25] All genes required for these pathways are found in *K. stuttgartiensis*; however, a gene encoding ATP citrate lyase is absent, which indicates that the citric acid cycle was mainly used for amino acid metabolism. The reducing equivalents required for CO_2 reduction are derived from the oxidation of reduced quinone or NADH. Clusters coding for NADH:ubiquinone oxidoreductase (complex I), a sodium-translocating NADH:quinone oxidoreductase and a formate:quinone oxidoreductase complex were identified in the genome and this complies with such electron transfer. Moreover, the genome of *K. stuttgartiensis* showed the presence of a versatile lifestyle[9] and around 200 genes were predicted to be directly involved in respiration.[24] The carbon metabolism was reversible and several transporters related to organic acid and amino acid metabolism were found. Moreover, three gene clusters encoding complex III (cytochrome bc1) and four gene clusters coding for ATP synthase complexes were found, which proved that ATP is synthesized by a chemiosmotic mechanism. Ladderanes and their derivatives are of great importance in anammox bacteria; however, the pathway of ladderane biosynthesis is still unknown. The genome of *K. stuttgartiensis* shows the presence of biosynthesis pathways for the formation of fatty acids, squalene, hopanoids and ubiquinones.[24] Apart from them, genes and operons for *S*-adenosyl methionine (SAM) methylases, putative radical SAM enzymes and a gene similar to phenylacetate-coenzyme A ligase were also identified. Moreover, anammox bacteria have the capacity to biosynthesize several unknown molecules that are involved in methylation, cyclization *via* oxidative radical chemistry and fatty acid elongation.

3.3 Ecological and Environmental Significance of Anammox Bacteria

Several methods, such as 16S rRNA gene-targeted phylogenetic analysis, have been used extensively to describe the organisms present in environmental samples. Later, with the advent of several metagenomics technologies, it has become easier to find unknown organisms from environmental samples. Several functional PCR approaches have been used to amplify genes related to encoding hydroxylamine/hydrazine oxidoreductase (HAO/HZO) proteins, which can serve as a more suitable target for ecological studies for both aerobic and anaerobic ammonium-oxidizing bacteria.[11] Fluorescence *in situ* hybridization (FISH) is also used as a tool to analyse environmental samples and used to validate the findings of clone libraries. FISH-MAR and ISR probing have been used to measure the activity and growth at the single-cell level.[8] Confocal Raman microscopy has also been used to identify anammox bacteria in a non-invasive way.[28] Tracer techniques like using ^{15}N-labeled ammonium and nitrite can also be used to detect any anammox activity in environmental samples.[29] Differential effects of acetylene and methanol on anammox and denitrification have helped to elucidate the contribution of these in the main pathways of N_2 production in marine sediments.[30] Moreover, ladderane lipids of anammox bacteria can also be used as biomarkers.

Anammox bacteria usually interact with different groups of bacteria and assist in the treatment of wastewater in wastewater treatment plants, which helps in metabolizing non-ammonium pollutants in partial nitritation/anammox processes (Figure 3.1).

There remains the possibility of a competitive interaction between denitrifiers and anammox bacteria as they utilize the same electron acceptor, NO_2^-; however, synergistic relations between these two groups occur. Accumulation of NO_2^- can lead to toxicity to anammox bacteria,[31] and the removal of NO_2^- by the denitrifiers may preserve anammox activity at higher NO_2^- concentrations. Heterotrophic denitrifiers also mineralize organic nitrogen to ammonia that can be used by anammox bacteria, and also help in the removal of chemical oxygen demand (COD) and total nitrogen. The denitrifiers also reduce NO_3 formed by anammox to N_2 gas, which is released into the atmosphere or again used by anammox bacteria, which results in higher TN removal. Denitrification can also supply NO_2^- to anammox in natural wetlands and river sediments. Thus, for efficient removal of total nitrogen and COD, a stable interaction between denitrifiers and anammox bacteria is required. In pre-treated domestic sewage, anammox bacteria were dominant along with a minor population of heterotrophic denitrifiers for N-removal.[32,33] Moreover, it was seen that a low COD level had no impact on the anammox bacteria, but enhanced the total nitrogen removal by heterotrophic bacteria.[34]

Anaerobically treated sewage often contains dissolved methane (CH_4), which in oversaturated conditions becomes difficult to recover.[35] In the absence of any effective CH_4 collection system, the gas can be released into the air, which may create the risk of explosion and add to the global warming potential by as high as 34 CO_2 equivalents.[36] To date, the microbial consortium that oxidizes CH_4 to CO_2 coupled with NO_3^- reduction to N_2 has been enriched and a nitrate-dependent archaeon (Candidatus *Methanoperedens nitroreducens*) has been reported.[37] They interact with anammox bacteria in CH_4-laden anaerobically-treated sewage. Candidatus *Methanoperedens nitroreducens* has been reported to reduce NO_3^- to NO_2^-, which is further

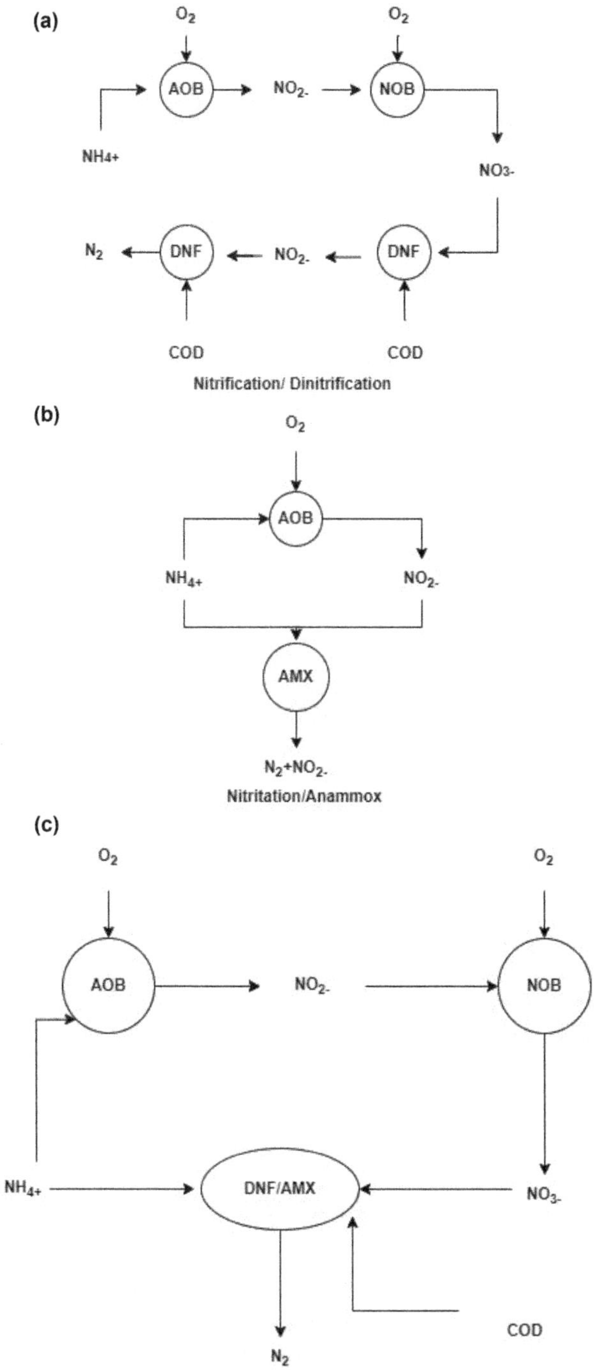

Figure 3.1 Nitrogen conversion in (a) nitrification/denitrification, (b) nitritation/anammox and (c) denitritation/anammox [AOB = ammonium oxidizing bacteria; NOB = nitrite oxidizing bacteria; AMX = anammox bacteria; DNF = denitrifier].

reduced to N_2 by anammox bacteria. It was also found that *Candidatus Methylomirabilis oxyfera*, which reduces NO_2^- to N_2 gas, competes with the anammox bacteria for NO_2. Thus, the coexistence of nitrate/nitrite-dependent anaerobic methane oxidizing bacteria and anammox bacteria plays a crucial role in the complete removal of CH_4 and TN from anaerobically-treated sewage.[38] Winkler *et al.*[39] developed a mathematical model on the coexistence of nitrite-dependent AMO bacteria and anammox bacteria which showed that high NO_2^-/NH_4^+ ratio and biomass concentration are crucial for the achievement of NH_4^+ and CH_4 removal. Luesken *et al.*[40] reported an enriched culture of *Candidatus Methylomirabilis oxyfera* and anammox bacteria, such as *Candidatus Jettenia asiatica* and *Candidatus Brocadia fulgida*, in the presence of limited NH_4^+ and excess CH_4 and NO_2. It was found that *Candidatus Methylomirabilis oxyfera* competes with the anammox bacteria for NO_2^-; however, as the affinity of anammox bacteria towards NO_2^- is higher, it results in more NO_2^- consumption. Haroon *et al.*[37] also reported that a partnership between *Candidatus Methanoperedens nitroreducens* and anammox bacteria can effectively outcompete *Candidatus Methylomirabilis oxyfera*. However, insights into how environmental factors regulate the interaction between these bacterial communities are still not clear and further investigations are urgently required to provide guidance for the regulation of the operational parameters.

Anammox bacteria were also found to be stimulated by the addition of sulfide, sulfur, sulfite and thiosulfate[7] as with these reduced S-compounds there is an extra provision of NO_2^-. The coexistence of anammox and NO_3^- aiding chemolithotrophic oxidation of sulfide to colloidal sulfur and/or sulfate[41] was also evident. The process of oxidation of sulfide, and then subsequently to sulfur and sulfate has been seen depending on the availability of an electron acceptor. Sulfide can stimulate anammox activity, but can also exert substrate-dependent inhibition, which was highly dependent on the substrate concentration and exposure time.[42] Dapena-Mora *et al.*[43] reported that 1–2 mM of sulfide suppresses anammox activity by 60%. A coculture of anammox bacteria and autotrophic denitrifiers such as *Sedimenticola* sp., which consumes sulfide, NH_4^+, NO_3^- and CO_2, also supported this theory.[44] The presence of reduced sulfur compounds in anaerobically treated sewage facilitates the reduction of NO_3^- to NO_2^- by autotrophic denitrifiers, leading to enhanced total nitrogen.

3.4 Industrial Application of the Anammox Process

Anammox organisms are detected and identified in wastewater treatment plants and in marine and fresh-water habitats. Ammonia being a hazardous nitrogenous compound has severe health implications and several biological approaches have been utilized to remove it from wastewater. It is reported that microalgae-based assimilation, non-thermal plasma coupled with natural minerals like zeolites, membrane technology and heterotrophic ammonia oxidation have been previously employed in ammonia removal.[45]

It was realized by researchers that anammox bacteria have huge potential for the removal of excess ammonium from wastewater systems and Mulder *et al.*[5] patented the process. The conventional nitrogen removal largely comprises nitrification and denitrification processes, where nitrification is a bi-step process catalyzed by obligate chemolithoautotrophic prokaryotes including nitrite-oxidizing bacteria (NOB) and ammonia-oxidizing archaea (AOA) or ammonia-oxidizing bacteria (AOB). Both ammonia-oxidizing archaea (AOA) and ammonia-oxidizing bacteria (AOB) are known to oxidize ammonia to nitrite, and later nitrite-oxidizing bacteria (NOB) oxidize nitrite to form nitrate. All of these bacterial and archaeal groups are aerobic and chemolithoautotrophic, and they fix CO_2 and gain energy for growth by oxidizing ammonia or nitrite.[46] On the other hand, denitrification is the final step, which is an anaerobic process catalyzed by facultatively anaerobic heterotrophic bacteria, which help in nitrogen removal. In each case, the bacteria use organic compounds as electron donors for reducing NO_3^- and NO_2^- to their gaseous forms, such as NO, N_2O, and N_2.[47] Later Van Dongen *et al.*[48] devised a large-scale single reactor system for high-rate ammonium removal over nitrite (SHARON) where ammonium is converted to nitrite, rather than nitrate. The inoculum mixture was produced in the laboratory in an aerated, stirred tank reactor under controlled conditions, where the nitrite oxidizers, such as *Nitrobacter* spp., are washed out. As wastewater was fed into the reactor, there was a complete conversion of the ammonium to N_2 gas and additional nitrate. This phenomenon when applied to a large-scale plant showed a start-up period of approximately 2 years due to several factors, such as low growth rates of anammox microorganisms, and other technical issues. Initially the reactor was fed with nitrifying sludge and during full operation showed a conversion between 8–10 kg of nitrogen per m^3 every day. A third large-scale, oxygen-limited reactor that functions on the principle of the CANON process is also in operation. Liu *et al.*[49] reported that anammox bacteria have a high tolerance towards free ammonia when compared with that of microalgae-based ammonia treatment and also the inhibition created on anammox bacteria is irreversible. Moreover, the operational cost is much lower and the efficiency of the treatment is not jeopardised by increasing the ammonia level as is seen in non-thermal plasma coupled with zeolite treatment mechanisms.[50] It was reported that heterotrophic ammonia-oxidizing bacteria (HAOB) can be used for the removal of ammonia;[45] however, they are inhibited by heavy metals such as Cu(II) and are suspectable to lower temperatures.[45] Meanwhile, through the anammox process, a higher level of nitrogen removal can be achieved despite of factors such as low oxygen availability, no external carbon source and no carbon dioxide emission.[51] The anammox bacterial load is tolerant to ammonium and nitrogen,[4] organic waste and fluorine and phenol and thus considered to be more efficient in wastewater treatment. Thus, nitritation/anammox processes have become very popular in wastewater treatment systems. However, this process of anammox and partial nitritation (PN)/anammox (PN-A) is largely applied for the treatment of warm effluents having a temperature ranging from 20 °C to 35 °C and operating this process at low temperature is challenging. The major drawbacks of this process are lower nitrogen removal efficiency, long-term instability in the overall process and production of nitrate as one of the end products, which needs to be addressed to broaden the applications of the anammox process. The anammox process aided by partial denitrification could be a good alternative and could be a successful venture to treat wastewater. Partial denitrification generates nitrite from nitrates and has specificity towards nitrates.[47] Zhang *et al.*[52] reported that a combination of the anammox process and partial denitrification can achieve a total nitrogen removal efficiency of more than 90%. Furthermore, increasing the pH in the denitrification stops the generation of N_2O, which is a greenhouse gas, and the lower sensitivity towards environmental factors provides efficient nitrogen removal from wastewater. As complete removal of nitrogen requires a huge amount of organic carbon and biofilm formation is considered as a promising solution for anammox biomass retention, a biofilm-based hybrid system has been suggested and demonstrated by many research groups. The process of nitritation along with anammox together with pretreatment of the sludge for organic removal can also serve as an eco-friendly cost-effective strategy for wastewater treatment.[46] However, it is essential to maintain optimum operational conditions, which is a challenging task. The main drawbacks of the process include long start-up period, low nitrogen concentration and inconsistency in the loading rates.[53] This process of wastewater treatment can be applied to wastewater containing a wide range of pollutants, such as pharmaceutical wastewater, landfill leachate, and swine wastewater on industrial scales; however, future research is necessary to improve its efficiency.

Methane-based reactors along with anammox and n-DAMO microorganisms can also be used for high level nitrogen removal and in this case a dominance of anammox bacteria was reported.[54] The authors predicted the formation of biofilms attributed to the microbial yield of anammox bacteria. The higher anammox bacteria population over n-DAMO bacteria and n-DAMO archaea was predicted on the basis of thermodynamic analysis. Li *et al.*[55] reported higher removal of total nitrogen by autotrophic denitrification and the anammox process in wastewater containing fluorine; similarly, higher removal efficiency of ammonia (97.6%) was also achieved in low C/N domestic wastewater at low temperatures.[56] It was proven that a novel single-stage process integrating simultaneous COD oxidation, partial nitritation–denitritation, and anammox (SCONDA) could achieve a high COD removal.[57] Fan *et al.*[4] also reported efficient removal of ammonium from wastewater by a two stage anammox process, which was characterized by a reduced hydraulic retention time and gradual increase in species richness diversity within reactors. *Candidatus Kuenenia* was observed as the dominant anammox bacterium and all findings of this study provide new insights into the potential application of the two-stage anammox process.

3.4.1 Factors Influencing the Anammox Bacteria

The maintenance of the anammox systems to attain successful removal of nitrogen from wastewater is quite challenging. The process can be inhibited by a diverse range of pollutants such as heavy metals, antibiotics, phenols and alcohol which are common in wastewater. Moreover, the concentration of organic compounds, pH, oxygen and nitrite is also vital for the maintenance of the anammox process.[31] Similarly, other bacterial communities thriving in the wastewater also limit the activity growth and the growth of anammox bacteria.

3.4.1.1 Effect of Heavy Metals

Industrial wastewater is contaminated with toxic heavy metals, which are toxic to most microorganisms resulting in cell death, leading to cell growth inhibition.[58] The most common heavy metals present in the wastewater include Cu, Cr, Pb, Zn, Ni, Cd, and Mn. Copper is an essential micronutrient required by living organisms, but it is toxic at high concentrations (9 to 19.3 $mg\,L^{-1}$) and the toxic form is Cu(II).[58] Similarly, zinc becomes toxic beyond 3.9–59.1 $mg\,L^{-1}$ and its increase leads to a 50% reduction in anammox activity in a treatment plant. Cadmium is also a toxic metal that inhibits the anammox process; however, it was found that *Candidatus Brocadia fulgida* showed higher tolerance to that metal.[58] Anammox granules were more resilient towards Pb than anammox biofilms.[59] Similarly, other toxic metals such as Hg, Ag, Cr, Mo, Co, and As also demonstrated toxicity affecting anammox activity[58] (Cho *et al.* 2020). According to Kimura and Isaka,[60] the inhibitory effect of Cu, Ni, Co, and Zn on anammox activity was observed to be reversible, while that from Mo was shown to be irreversible. Past research studies have shown that anammox bacteria accumulate heavy metals, which are accumulated in extracellular polymeric substances and prevent the activity of the anammox process.

3.4.1.2 Effect of Organic Substances

Anammox bacteria are chemoautotrophic in nature and organic carbon interferes with the growth of anammox bacteria. The inhibition of anammox bacteria occurs due to outcompetition and fast growth of heterotrophic bacteria in the presence of organic compounds.[61] Toxic organic compounds such as alcohol, phenol, aldehydes, and antibiotics inhibit the activity of anammox bacteria irreversibly, but the mechanism of toxicity is not clear. *Candidatus 'Anammoxoglobus propionicus'* has been observed to compete with other anammox bacteria and heterotrophs when exposed to propionate as a secondary carbon source.

Ammonia is the primary substrate of the anammox process. Some wastewater streams also contain monosodium glutamate wastewater, landfill leachate, wastewater from nitrogen fertilizer plants, coking wastewater, and other industrial wastewater, which are rich in ammonia making them suitable for the anammox process. Although, if the ammonium exceeds 1 gN L^{-1} it inhibits the anammox process.[62] Free ammonia is more inhibitory than ammonium to anammox bacteria.[49] As anammox bacteria are known to form biofilms, the availability of free ammonia is impacted. Ammonia (NH_3) is able to diffuse across a cell membrane, whereas ammonium (NH_4^+) cannot diffuse freely.[49,62] The inhibitory impact of NH_3 is due to it changing the intracellular pH, which alters the transmembrane potential.[62]

Although, nitrite is an electron acceptor required by anammox bacteria, the biotoxicity of nitrite is also reported and only a threshold limit is applicable for the bacterial activity.[58] However, it is reported that, in comparison to heterotrophic denitrifying bacteria, anammox bacteria are capable of tolerating higher nitrite concentrations.[63] Furthermore, when comparing nitrite toxicity with free ammonia toxicity, anammox was found to be more vulnerable to nitrite than free ammonia. The impact of nitrite is more inhibitory in nature, rather than toxic, which was mainly due to the change in pH.

3.4.1.3 *Effect of Temperature*

The effect of temperature has a strong impact on the growth of anammox bacteria by affecting nitrogen removal, activating inhibitory factors that are detrimental to the microbial community, and enhancing the stability of organic chemical effluent water.[46] Anammox bacteria were found to be tolerant of a wide range of temperatures ranging from 4–80 °C; however, the best tolerable temperatures range between 20–45 °C.[64] Earlier, a low temperature around 15 °C was preferred for efficient removal of nitrogen by the anammox process, but later it was found that lower temperature results in lower removal of nitrogen and tends to accumulate nitrogen as nitrate in the wastewater. Usually isolates like *Candidatus brocadia* are able to survive under cold conditions.[65] Moreover, at low temperature, more ladderane fatty acids are present in the membrane of the anammoxosome, which reduces the membrane permeability and prevents ATP synthesis.[66] Jin *et al.*[62] reported that higher temperature 30–40 °C anammox growth reduces the emission of greenhouse gases such as CO_2 and N_2O. A temperature above 40 °C is detrimental to anammox bacteria, mainly due to the cellular lysis and the release of cytochrome out from the cells.

3.4.1.4 *Effect of pH*

It was reported that the most suitable pH for anammox activity ranged from 6.5 to 9.0.[64] A change in pH results in the formation of toxic chemicals such as free nitrous acid, which is inhibitory in nature.[58] The tolerance towards pH depends on the type of biomass, nature of the aggregates, nature of biofilm formation and anammox bacteria species. *Candidatus Brocadia anammoxidans* and *Candidatus Kuenenia stuttgartiensis* dominate at pH between 8 and 8.5,

Table 3.1 Factors influencing anammox activity.[68]

Conditions	Threshold condition for inhibition	Effect	Ref.
Temperature	13–18 °C	Reduced growth rate	69
pH	7	Reduced growth rate	69
Nitrite	53 $mg\,N\,L^{-1}$	Reduction in the anammox activity	70
Sulfides	320 $mg\,L^{-1}$	Reduction in the anammox activity	71
Salinity	NaCl: 13.45 $mg\,L^{-1}$, Na_2SO_4: 11.36 $mg\,L^{-1}$, KCl: 14.9	Reduction in the anammox activity	43
Dissolved oxygen	$<$0.04 $mg\,L^{-1}$	Reduction in the anammox activity	72

whereas *Candidatus Brocadia anammoxidans* and *Candidatus Anammoxoglobus propionicus* dominate at near neutral pH.[11,67]

Detailed information on the factors that are responsible for anammox activity is given in Table 3.1.

3.5 Future Prospects

The limited knowledge on anammox metabolism is considered as the main drawback that has restricted its efficient use in wastewater treatment. Moreover, no clear information is available on the interaction of anammox bacteria with other bacterial groups associated with nitrogen cycles. Also, it is difficult to produce a pure culture of anammox bacteria, which has prevented detailed studies of its metabolic pathways, physiological properties, biochemical characteristics, and genetics. The functional dynamics and the composition of microbial groups, including anammox bacteria, in a PN-A system affects the nitrogen removal rate from wastewater treatment systems. Therefore, further studies on microbial community composition in anammox reactor systems are the need of the time. Both metagenomic and metatranscriptomic analyses can be successfully used to understand the gene expression, enzyme activity and metabolic capabilities of uncultured anammox species under different environmental conditions. Also, further studies focusing on the performance of anammox systems under varying seasonal conditions need to be done to increase the efficiency of the anammox process in wastewater treatment. The main limitations, including extended start-up periods, long doubling period and lower biomass yield of anammox bacteria, need to be overcome. Genetic alteration using different biotechnological processes can act as a good option to increase the efficiency of anammox bacteria; however, for that natural genetic and metabolic diversity should be first exploited. Moreover, anammox bacteria are known to form biofilms within the reactor, and maintenance of the biofilm is also crucial for a successful biomass yield of anammox cells. Similarly, up-to-date knowledge on diverse methods and different reactor systems to improve the nitrogen removal efficiency should be gained to improve wastewater treatment. Considering the cost-effective advantages of the anammox process to remove ammonium, strategies should be developed for its comprehensive application for treating both municipal and industrial waste streams in future.

References

1. A. Hasanoğlu, J. Romero, B. Pérez and A. Plaza, *Chem. Eng. J.*, 2010, **160**(2), 530–537.
2. R. R. Karri, J. N. Sahu and V. Chimmiri, *J. Mol. Liq.*, 2018, **261**, 21–31.
3. P. Seruga, M. Krzywonos, J. Pyzanowska, A. Urbanowska, H. Pawlak-Kruczek and L. Niedźwiecki, *Molecules*, 2019, **24**(20), 3633.
4. N.-S. Fan, Y.-H. Bai, J. Wu, Q. Zhang, J.-J. Fu, W.-L. Zhou, B.-C. Huang and R.-C. Jin, *J. Cleaner Prod.*, 2020, **261**, 121148.
5. A. Mulder, A. A. Graaf, L. A. Robertson and J. G. Kuenen, *FEMS Microbiol. Ecol.*, 1995, **16**(3), 177–183.
6. S. Li, C. Kambey and V. Nguyen, *Water*, 2020, **12**(1), 20.
7. A. A. van de Graaf, P. de Bruijn, L. A. Robertson, M. S. M. Jetten and J. G. Kuenen, *Microbiology*, 1996, **142**, 2187–2196.
8. M. Schmid, S. Schmitz-Esser, M. Jetten and M. Wagner, *Environ. Microbiol.*, 2001, **3**, 450–459.
9. B. Kartal, L. van Niftrik, J. Rattray, J. de Vossenberg, M. C. Schmid, J. S. S. Damste and M. Strous, *FEMS Microbiol. Ecol.*, 2008, **63**, 46–55.
10. B. Kartal, M. M. M. Kuypers, G. Lavik, J. Schalk, H. J. M. Op den Camp, M. S. M. Jetten and M. Strous, *Environ. Microbiol.*, 2007, **9**, 635–642.
11. Z. Quan, S. Rhee, J. Zuo, Y. Yang, J. Bae, J. R. Park, S. Lee and Y. Park, *Environ. Microbiol.*, 2008, **10**, 3130–3139.
12. A. Rodriguez-Sanchez, J. Purswani, T. Lotti, P. Maza-Marquez, M. C. M. Van Loosdrecht, R. Vahala and A. Gonzalez-Martinez, *Environ. Technol.*, 2016, **37**(18), 2281–2291.
13. M. Oshiki, M. Ali, K. Shinyako-Hata, H. Satoh and S. Okabe, *Environ. Microbiol.*, 2016, **18**(9), 3133–3143.
14. P. Han and J. D. Gu, *Geomicrobiol. J.*, 2015, **32**(9), 789–798.
15. M. M. Kuypers, A. O. Sliekers, G. Lavik, M. Schmid, B. B. Jørgensen, J. G. Kuenen, J. S. Sinninghe Damsté, M. Strous and M. S. Jetten, *Nature*, 2003, **422**(6932), 608–611. PMID: 12686999.
16. J. A. Fuerst, *Annu. Rev. Microbiol.*, 2005, **59**, 299–328.
17. M. R. Lindsay, R. I. Webb, M. Strous, M. S. M. Jetten, M. K. Butler, R. J. Forde and J. A. Fuerst, *Arch. Microbiol.*, 2001, **175**, 413–429.
18. L. van Niftrik, W. J. C. Geerts, E. G. van Donselaar, B. M. Humbel, R. I. Webb, J. A. Fuerst, A. J. Verkleij, M. S. M. Jetten and M. Strous, *J. Bacteriol.*, 2008, **190**, 708–717.
19. M. Ali, D. R. Shaw, L. Zhang, M. F. Haroon, Y. Narita, A. H. Emwas and S. Okabe, *Water Res.*, 2018, **143**, 10–18.

20. M. Ali, M. F. Haroon, Y. Narita, L. Zhang, D. R. Shaw, S. Okabe and P. E. Saikaly, *Genome Announce.*, 2016, **4**, e01377-16.
21. D. G. Weissbrodt, N. Shani, L. Sinclair, G. Lefebvre, P. Rossi and J. Maillard, *et al.*, *BMC Microbiol.*, 2012, **12**, 306.
22. L. Zhang, Y. Narita, L. Gao, M. Ali, M. Oshiki, S. Ishii and S. Okabe, *Water Res.*, 2017, **125**, 249–258.
23. M. Strous, E. Van Gerven, J. G. Kuenen and M. Jetten, *Appl. Environ. Microbiol.*, 1997, **63**, 2446–2448.
24. M. Strous, E. Pelletier, S. Mangenot, T. Rattei, A. Lehner, M. W. Taylor and D. Le Paslier, *Nature*, 2006, **440**, 790–794.
25. M. Strous, J. J. Heijnen, J. G. Kuenen and M. S. M. Jetten, *Environ. Sci. Technol.*, 1998, **50**, 589–596.
26. L. L. Straka, K. A. Meinhardt, A. Bollmann, D. A. Stahl and M. K. Winkler, *ISME J.*, 2019, **13**(8), 1997–2004.
27. L. Chistoserdova, C. Jenkins, M. G. Kalyuzhnaya, C. J. Marx, A. L. Julia, A. V. James, T. Staley and M. E. Lidstrom, *Mol. Biol. Evol.*, 2004, **21**(7), 1234–1241.
28. R. Pätzold, M. Keuntje, K. Theophile, J. Müller, E. Mielcarek, A. Ngezahayo and A. Anders-von Ahlften, *J. Microbiol. Methods*, 2008, **72**, 241–248.
29. N. Risgaard-Petersen, L. P. Nielsen, S. Rysgaard, T. Dalsgaard and R. L. Meyer, *Limnol. Oceanogr.: Methods*, 2003, **1**, 63–73.
30. M. M. Jensen, B. Thamdrup and T. Dalsgaard, *Appl. Environ. Microbiol.*, 2007, **73**, 3151–3158.
31. R. C. Cho, G. F. Yang, J. J. Yu and P. Zheng, *Chem. Eng. J.*, 2012, **197**, 67–79.
32. M. Laureni, D. G. Weissbrodt, I. Szivak, O. Robin, J. L. Nielsen, E. Morgenroth and A. Joss, *Water Res.*, 2015, **80**, 325–336.
33. T. Lotti, R. Kleerebezem, Z. Hu, B. Kartal, M. S. M. Jetten and M. C. M. van Loosdrecht, *Water Res.*, 2014, **66**, 111–121.
34. S. Q. Ni, J. Y. Ni, D. L. Hu and S. Sung, *Bioresour. Technol.*, 2012, **110**, 701–705.
35. T. L. Hendrickx, Y. Wang, C. Kampman, G. Zeeman, H. Temmink and C. J. Buisman, *Water Res.*, 2012, **46**, 2187–2193.
36. *Climate change 2013: The physical science basis. Contribution of Working Group I to the fifth assessment report of the intergovernmental panel on climate change*, ed. T. F. Stocker, D. Qin, G. K. Plattner, M. Tignor, S. K. Allen, J. Boschung and P. M. Midgley, Cambridge University Press, 2013.
37. M. F. Haroon, S. H. Hu, Y. Shi, M. Imelfort, J. Keller, P. Hugenholtz and G. W. Tyson, *Nature*, 2013, **500**, 567–570.
38. K. F. Ettwig, B. L. Zhu, D. Speth, J. T. Keltjens, M. S. M. Jetten and B. Kartal, *Proc. Natl. Acad. Sci. U. S. A.*, 2016, **113**, 12792–12796.
39. M. H. Winkler, K. Ettwig, T. Vannecke, K. Stultiens, A. Bogdan, B. Kartal and E. Volcke, *Water Res.*, 2015, **73**, 323–331.
40. F. A. Luesken, J. Sanchez, T. A. van Alen, J. Sanabria, H. J. M. Op den Camp, M. S. M. Jetten and B. Kartal, *Appl. Environ. Microbiol.*, 2011, **77**, 6802–6807.

41. C. B. Wenk, J. Blees, J. Zopfi, M. Veronesi, A. Bourbonnais, C. J. Schubert, H. Niemann and M. F. Lehmann, *Limnol. Oceanogr.*, 2013, **58**(1), 1–2.
42. R. C. Jin, G. F. Yang, Q. Q. Zhang, C. Ma, J. J. Yu and B. S. Xing, *Water Res.*, 2013, **47**, 1459–1469.
43. A. Dapena-Mora, I. Fernandez, J. L. Campos, A. Mosquera-Corral, R. Mendez and M. S. M. Jetten, *Enzyme Microb. Technol.*, 2007, **40**, 859–865.
44. L. Russ, D. R. Speth, M. S. M. Jetten, H. J. M. Op den Camp and B. Kartal, *Environ. Microbiol.*, 2014, **16**, 3487–3498.
45. Z. Yin, X. Yang, W. Xiaolong, W. Xingyu and S. Wang, *Bioresour. Technol.*, 2021, **328**, 124823.
46. J. Li, J. Li, R. Gao, M. Wang, L. Yang, X. Wang, L. Zhang and Y. Peng, *Bioresour. Technol.*, 2018, **265**, 498–505.
47. R. Du, Y. Peng, J. Ji, L. Shi, R. Gao and X. Li, *Environ. Int.*, 2019, **131**, 105001.
48. U. G. J. M. Van Dongen, M. S. M. Jetten and M. van Loosdrecht, *Water Sci. Technol.*, 2001, **44**(1), 153–160.
49. Y. Liu, H. H. Ngo, W. Guo, L. Peng, D. Wang and B. Ni, *Environ. Int.*, 2019, **123**, 10–19.
50. J. Fan, H. Wu, R. Liu, L. Meng, Z. Fang, F. Liu and Y. Xu, *J. Hazard. Mater.*, 2021, **401**, 123627.
51. X.-Y. Jiang, Y.-F. Cheng, W.-Q. Zhu, Y.-H. Bai, L.-Z.-J. Xu, X.-Q. Wu and R.-C. Jin, *Bioresour. Technol.*, 2018, **260**, 1–8.
52. W. Zhang, Y. Peng, L. Zhang, X. Li and Q. Zhang, *Bioresour. Technol.*, 2020, **305**, 123045.
53. G. Zhu, S. Wang, B. Ma, X. Wang, J. Zhou, S. Zhao and R. Liu, *Water Res.*, 2018, **142**, 147–158.
54. T. Liu, J. Guo, S. Hu and Z. Yuan, *Environ. Int.*, 2020, **137**, 105501.
55. X. Li, Y. Yuan, Y. Huang and Z. Bi, *Sci. Total Environ.*, 2019, **661**, 235–242.
56. R. Du, S. Cao, S. Wang, M. Niu and Y. Peng, *Bioresour. Technol.*, 2019, **219**, 420–429.
57. X. Zhou, X. Zhang, Z. Zhang and Y. Liu, *Bioresour. Technol.*, 2018, **261**, 379–384.
58. S. Cho, C. Kambey and V. Nguyen, *Water*, 2020, **12**(1), 20.
59. G. Li, D. Puyol, J. M. Carvajal-Arroyo, R. Sierra-Alvarez and J. A. Field, *J. Chem. Technol. Biotechnol.*, 2015, **90**(5), 830–837.
60. Y. Kimura and K. Isaka, *Appl. Microbiol. Biotechnol.*, 2014, **98**(16), 6965–6972.
61. M. R. Mattei, L. Frunzo, B. D'Acunto, G. Esposito and F. Pirozzi, *Ecol. Model.*, 2015, **304**, 44–58.
62. R. C. Jin, G. F. Yang, J. J. Yu and P. Zheng, *Chem. Eng. J.*, 2012, **197**, 67–79.
63. Y. I. Ren, H. Hao Ngo, W. Guo, D. Wang, L. Peng, B.-J. Ni, W. Wei and Y. Liu, *Bioresour. Technol.*, 2020, **297**, 122491.
64. A. D. Pereira, A. Cabezas, C. Etchebehere, C. A. D. Chernicharo and J. C. de Araújo, *Environ. Technol. Rev.*, 2017, **6**(1), 74–93.

65. T. Lotti, R. Kleerebezem, Z. Hu, B. Kartal, M. K. de Kreuk, C. V. E. T. Kip and M. C. M. van Loosdrecht, *Environ. Technol.*, 2015, **36**, 1167–1177.
66. F. R. Moss, S. R. Shuken, J. A. M. Mercer, C. M. Cohen, T. M. Weiss, S. G. Boxer and N. Z. Burns, *Proc. Natl. Acad. Sci. U. S. A.*, 2018, **115**(37), 9098–9103.
67. S. He, Q. Niu, H. Ma, Y. Zhang and Y. Y. Li, *Water, Air, Soil Pollut.*, 2015, **226**(5), 163.
68. C. Weralupitiya, R. Wanigatunge, S. Joseph, B. Athapattu, T. H. Lee, J. K. Biswas and M. Vithanage, *Bioresour. Technol.*, 2021, 125240.
69. S. He, Y. Chen, M. Qin, Z. Mao, L. Yuan, Q. Niu and X. Tan, *Bioresour. Technol.*, 2018, **260**, 186–195.
70. J. M. Carvajal-Arroyo, D. Puyol, G. Li, A. Lucero-Acuña, R. Sierra-Álvare and J. A. Field, *Water Res.*, 2014, **48**, 52–60.
71. J. M. Carvajal-Arroyo, W. Sun, R. Sierra-Alvarez and J. A. Field, *Chemosphere*, 2013, **91**(1), 22–27.
72. S. Liu, F. Yang, Y. Xue, Z. Gong, H. Chen, T. Wang and Z. Su, *Bioresour. Technol.*, 2008, **99**(17), 8273–8279.

CHAPTER 4

The Anaerobic Ammonium Oxidation Process: Inhibition, Challenges and Opportunities

PRANJAL P. DAS,* SIMONS DHARA AND MIHIR K. PURKAIT

Department of Chemical Engineering, Indian Institute of Technology Guwahati, Assam, India
*Email: pranj176107027@iitg.ac.in

4.1 Introduction

In general, wastewater treatments based on biological routes rely on microorganism's metabolic capability. They provide a more sustainable performance and lower operational expenses as compared to physical and chemical-based treatments. Eutrophication is caused by elements like phosphorus and nitrogen in the water, which reduce the quality of home and industrial water sources. The necessity for nitrogen elimination processes, particularly for organic effluents with higher ammonia levels, has increased. As a result, most treatment plants employ biological paths for nitrogen removal, based on nitrification and denitrification stages. However, high running expenses (mainly due to aeration), the formation of secondary contaminants, and the necessity to feed carbon resources limit these technologies significantly. Therefore, research based on waste water treatment is significantly focused on an anaerobic ammonium oxidation technique (anammox) to alleviate the above-described limitations.[1] The anammox method has been viewed as a promising technology for treating nitrogen wastewater with a high concentration as it creates less sludge and does not require additional carbon sources or aeration stages. Currently, the Changi

Ammonia Oxidizing Bacteria: Applications in Industrial Wastewater Treatment
Edited by Maulin P. Shah

Published by the Royal Society of Chemistry, www.rsc.org

water reclamation plant in Singapore is housing the largest anammox effluent plant with a capacity of 800 000 m^3 per day. Anammox is a cost-effective and energy-saving process, particularly related to aeration (0.9 $kW\,h\,kg^{-1}$ N), as compared to traditional nitrification–denitrification approaches (3.5 $kW\,h\,kg^{-1}$ N). In this process, a negligible amount of carbon source is required (0 kg of CH_3OH kg^{-1} N) for nitrogen removal as compared to conventional processes (3 kg of CH_3OH kg^{-1} N).[2] Furthermore, the energy requirement for oxygen supply in aeration is decreased by 63%. As a result, in the case of NH_4^+-rich effluent treatment, anammox processes have become more viable study topics, with potential for a wide range of growing applications. Anaerobic ammonium-oxidizing bacteria (AnAOB) are chemolithoautotrophs in nature. They can use NH_4^+ as a primary electron donor and nitrite (NO_2^-) ions as a terminal electron acceptor, respectively. Anammox processes are usually carried out in two steps. In the first step, partial nitritation converts half of the ammonium (NH_4^+) to nitrite under aerobic environments *via* the metabolism of mainly AOB and autotrophic ammonium-oxidizing microbes. *Nitrosomonas eutropha*, *Nitrosospira*, and *N. europaea* are examples of such microbes. In the latter step, anammox processes transform nitrite and ammonium into nitrogen gas in anaerobic circumstances, owing to the activity of anammox bacteria. These bacteria are typically Gram-negative and have a high amount of lipid in their structure. These anammox bacteria turn red due to their particular cytochrome during their enrichment in the culture.[3] Although anammox techniques have been widely implemented for sidestream wastewater treatment around the world, they have limited full-scale applicability in the case of mainstream effluent treatment. A range of components in the water, such as high NO_2^- content, sulfides, excessive salinity, toxic metals, and organic compounds, can destroy anammox bacteria and reduce the nitrogen removal efficacy. On the other hand, high COD/N ratios may allow heterotrophic bacteria to surpass anammox microbes in conventional wastewater treatment. Thus, it is harder to maintain nitritation paths by the preferential retention of AOB against other microbes, particularly nitrite-oxidizing bacteria (NOB). Furthermore, a few barriers like effluent temperature, high NO_3^- content, pH, the existence of organic elements and effective retention of anammox bacteria in low nitrogen environments, pose threats to the process's applicability and efficiency.[4] As a result, this research topic necessitates more critical evaluation and in-depth research.

This chapter presents a comprehensive assessment of the most recent studies on anammox inhibition, covering the influences of emerging contaminants as well as the underlying mechanism of inhibition and viable anammox processes to treat nitrogenous effluents. This chapter describes the various challenges of the anammox process, factors effecting the anammox-based process, difficulties in the retention of anammox microbes, the effect of high carbon content on process performance, *etc.* In addition, this chapter provides applications of the anammox process for full-scale operation and its future prospects.

4.2 Overview of Primary Anammox-based Processes

The anammox process is divided into three primary categories, as shown in Figure 4.1. The reactors are used for both nitrite and ammonium-contaminated synthetic effluents, instead of any nitrile-producing stage. In some cases, the real waste containing nitrite or ammonium is also operated as described in Figure 4.1a. This type of procedure is described as a pure anammox process. In another strategy, anammox (PNA) and partial nitritation processes are employed (Figure 4.1b). In recent times, the combined process of anammox (PDA) and denitritation has also emerged for nitrate-containing effluent treatment (Figure 4.1c). Besides, the PNA process is further characterized as a one-stage or two-stage unit, depending on the existence of the partial nitritation process. Figure 4.2 shows a schematic diagram of the two-stage PNA process. The two-stage PNA process can be operated in multiple modes, such as one-pass or by-pass.[5] In one-pass mode, all the influent drifts into the partial nitration section, where only 57% of the ammonium in the influent is converted into nitrite, achieving a ratio of 1.3 between NO_2^-–N and NH_4^+–N. On the contrary, the influent stream is separated into two sub-streams in the by-pass mode. The stream having 57% of the influent is sent to the nitration section for the total oxidation of the ammonium into nitrite. The resulting nitrile stream is then merged with the other sub-stream to attain the NO_2^-–N/NH_4^+–N ratio of 1.3.[6] The nitration, pure anammox, and PNA processes are discussed in the following sections.

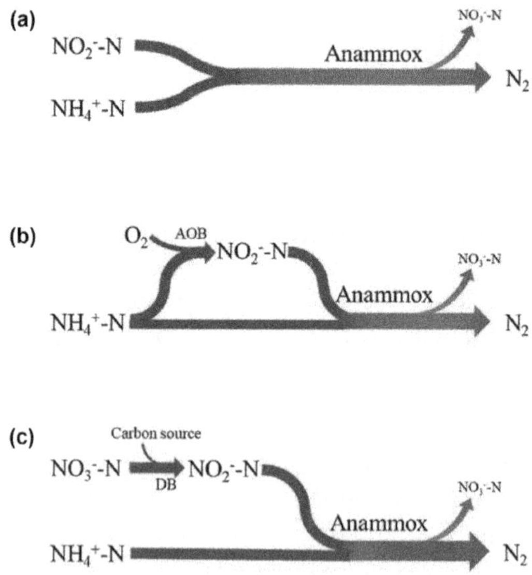

Figure 4.1 Three types of cases in anammox process research: (a) the pure anammox process, (b) the partial nitritation and anammox process; and (c) the denitritation and anammox process. Reproduced from ref. 5 with permission from Elsevier, Copyright 2020.

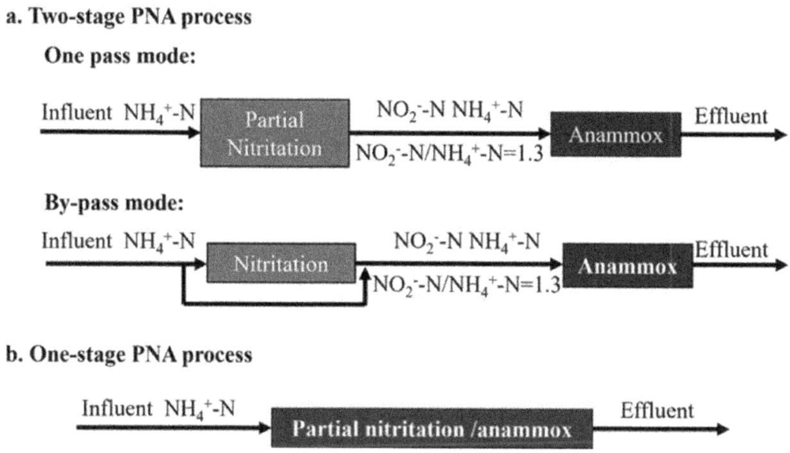

Figure 4.2 PNA process: (a) two-stage PNA process and (b) one-stage PNA process. Reproduced from ref. 5 with permission from Elsevier, Copyright 2020.

4.2.1 Nitritation Process

The majority of the biological nitrogen elimination techniques, such as the conventional nitrification–denitrification process, two-step PNA, and the nitrite shunt process, rely on the nitritation procedure to start up quickly and operate reliably. The procedure can be carried out by adjusting the temperature, substrate availability, pH, and free ammonia (FA) and free nitrous acid (FNA) inhibition levels to remove NOBs. Wenjie *et al.* (2016) reported the maximum nitrite accumulation rate (NAR) of 9.0–12.0 kg N per m^3 per day for rare-earth ammonium effluent treatment.[7] The NAR was significantly less for wastewater that had lower-ionic strength and lower-temperature conditions. In addition, a stable ratio of NO_2^-–N/NH_4^+–N was difficult to obtain in the case of the two-stage PNA process. Furthermore, in the case of a two-stage PNA procedure, achieving a consistent C/N ratio proved problematic.[8]

4.2.2 Pure Anammox Process

In the pure anammox process, it is feasible to achieve a very high nitrogen removal rate (NRR). Tang *et al.* (2011) found that the maximum NRR was 76.7 kg N per m^3 per day for the high-strength synthetic effluent.[9] On the contrary, the highest NRR was reported as 10.7 kg N per m^3 per day for a genuine waste brine solution. Therefore, the efficacy of the anammox process varies significantly between synthetic and actual wastewater. In contrast to high-strength wastewater, the value of the NRR was much lower for low-strength effluent, whether the effluent was synthetically generated or obtained as raw sewage. A significant amount of actual effluent is low-strength in nature, like waste collected from municipal sources. As a result, more research into enhancing the NRR for low-strength effluent is required. Recently, relatively high NRR was achieved with the help of the UASB configuration by the development of anammox granules and the build-up of bacteria in the anammox reactor.[10] In another example, high NRR was attained in the MBR reactor with the help of a high concentration of volatile suspended solids (VSS) by maintaining an extended solid retention time (SRT). A high NRR (1.51 kg N per m^3 per day) was also reported by Guo *et al.* (2018) in the anammox MBR configuration at a low strength of nitrogen.[11]

4.2.3 Partial Nitritation and Anammox Process

This process is mainly categorized as one-stage and two-stage PNA procedures. The two-stage PNA procedure has been largely preferred for raw wastewater handling. Although, according to research, the total NRR for the two-stage PNA is lower than that of the one-stage PNA. As a result, using the two-stage PNA approach to achieve a high NRR in the treatment of actual or raw wastewater is difficult. As per the survey conducted in 2014, one-stage PNA was used in over 80% of full-scale anammox-built systems.[12] Wang *et al.* (2017) reported a 73% NRE and a 7.1 kg N per m^3 per day NRR for an actual piggery effluent by using a one-stage PNA design.[13] The most common reactor configuration for one-stage PNA is the granular or biofilm sludge category, like the airlift reactor (ALR) or continuous stirred-tank reactor (CSTR). The outer surface of the granule or biofilm consists of ammonium oxidizers. This provides the anammox microorganisms with enough nitrite to survive in the interior layers.[5]

4.3 Factors Affecting the Anammox Process

4.3.1 Ammonia and Nitrite Concentrations

Anammox bacteria and AOB require the ammonium ion (NH_4^+), which is the major nitrogen component in the anaerobic digestion (AD) effluent. However, higher levels of NH_4^+ concentration cause anammox bacteria and AOB to become less active. In addition, anammox reactors have low alkalinity, so a certain quantity of NH_4^+ can mix with OH^- and create free ammonia (FA). FA suppresses the activity of microorganisms like AOB and NOB.[14] Additionally, NO_2^- formed in the partial nitrification step has an impact on the anammox process performance. It was found in one of the investigations that when the NO_2^- content reached up to 400 mg N per L, the reactor processing AD effluent failed. FNA derived from NO_2^- has also been found to have a devastating effect on AOB. The AOB activity was lowered by 50% with a 3-fold increment of FNA content (0.42–1.72 mg HNO_2^- N per L) in the anammox process. Therefore, it is essential to control the total nitrogen content in the influent of any anammox process. This can be accomplished by diluting the AD effluent through anammox process effluent recycling or by employing other appropriate technologies.[15]

4.3.2 Organic Concentration

As per the research conducted, a high COD/N ratio has a negative impact on the process performance for the treatment of AD effluent. The denitrification process will be the primary nitrogen elimination method when the COD loading rises. The reason behind this fact is that the activity of anammox bacteria can be influenced or stopped due to the presence of high organic carbon (OC). The activity of the anammox process was decreased in one investigation due to the presence of a COD level of 487.5 mg L^{-1} in the influent. In addition, too much OC content has a negative impact on the partial nitrification (PN) process, and sometimes the PN step can be inhibited if the OC level exceeds 2 kg per m^3 per day.[16] In the case of municipal sewage treatment, a COD level in the influent of up to 200 mg L^{-1} is advisable for a lab-based anammox process. The COD level can be raised appropriately for full-scale or pilot-based operations. The post digestion step is useful to reduce residual carbon in the influents of anammox processes. Thus, the OC loading can be decreased. The COD content of AD effluent is complicated, and it can be classified into three categories. These are rbCOD, sbCOD, and refractory organic compounds and are listed below.[17]

4.3.2.1 rbCOD

Anammox metabolism can be improved in the presence of extremely small amounts of volatile fatty acids such as acetate, formate, and propionate. However, according to Kindaichi *et al.* (2016), 50 mM of acetate in the influent reduced the inorganic carbon adsorption ability of anammox bacteria.[18]

4.3.2.2 sbCOD

One kind of sbCOD is represented as particulate settleable solids. AD effluent containing particulate solids at high levels raises the risk of blockage. Therefore, various pretreatment methods like filtration, coagulation, and interim storage can be used to minimize the solid concentrations in AD effluent before it enters into the anammox-based process.[19]

4.3.2.3 Refractory Organic Compounds

Recalcitrant OC does not easily effect the granular anammox sludge. However, the presence of refractory COD in the AD effluent is one of the primary sources of AOB inhibition in anammox-based systems.[20]

4.4 Inhibition of Wastewater Pollutants on Anammox Processes

4.4.1 Inhibition of Metals

Figure 4.3 shows the mechanism of pollutant inhibition in wastewater. Metals are often found in wastewater. Metals are necessary for AnAOB as trace elements. Several metal ions, including Cu^{2+}, Pb^{4+}, Zn^{2+}, Co^{2+}, Cr^{3+}, Cd^{2+}, Hg^{2+}, Mn^{2+}, Ni^{2+}, and, Fe^{2+} alter the anammox process, promoting it at low amounts and inhibiting it at higher concentration. Since it passes through the cell membrane and penetrates cells, damaging the enzyme configuration and reducing the cytochrome *c* quantity in the cell, Cd^{2+} displays persistent and acute suppression of the anammox mechanism. In a prior investigation, the IC_{50} of Cd^{2+} for short-term exposure was 5.43 mg L^{-1}, but the long-standing inhibition limit was 20 mg L^{-1}.[21] Hg^{2+} also lowers the efficiency of nitrogen removal and anammox process characteristics. Hg^{2+} has an IC_{50} of 60.39 mg L^{-1} and a degradation rate of nitrogen of 6.61 mg N per g VSS per h in sludge. The Hg^{2+} concentration was strongly associated with the presence of the Hg resistant genes. Co^{2+} boosted the anammox process' nitrogen removal effectiveness and raised the levels of essential enzymes in the cell at low doses. However, at high doses, it impeded the anammox process' activity. Cr^{6+} penetrated the cell and disrupted the enzyme structure, hindered the functional groups, and lowered the microorganisms' performance of nitrogen removal, according to previous research.[22] Furthermore, a small Fe^{3+} quantity (0.08 and 0.14 mmol L^{-1}) increased anammox sludge fractionation, with NLRs of 993.0 and 1105.8 kg N per m^3 per day for the anammox unit, respectively. In addition, a Fe^{2+} binding chelator aided cytochrome *c* production. Because of their stronger inhibition and bioavailabilities for AnAOB, Zn^{2+} and Cu^{2+} have been examined specifically. Zn^{2+} was shown to have antimicrobial effects and absorb UV rays. Zn^{2+} is a micronutrient that is required for the metalloproteinase cofactors and other enzymes. Large amounts of Zn^{2+} aggregate bind to proteins (PNs) in the cell engaged in the transportation of electrons.[23] In short-term experiments, the IC_{50} of Zn^{2+} was 6.9 ppm, whereas the lasting inhibition limit of Zn^{2+} on the anammox setup was 3.39 ppm. When the values of FA were 48.4–95.0 ppm in the anammox setup, the IC_{50} of Cu^{2+} was found to be 74 ppm. Once the feed water concentration was 280 ppm, the IC_{50} of Cu^{2+} was 1.9–30 ppm. Because of this disparity, it's likely that substrate circumstances altered the inhibitory effects of different contaminants.[24] In recent years, research has focused on the influence of mono metal factors on the embedded methods as well as on the cumulative effects of different contaminants and metals, like antibiotics and metals. Cu^{2+} bioavailability and suppression were shown to be reduced when it was coupled with antibiotics in a prior investigation. At present, the focus of metal research has turned away from anammox inhibition and toward genome analysis.[25] The positive and negative effects of different metals are shown in Table 4.1.

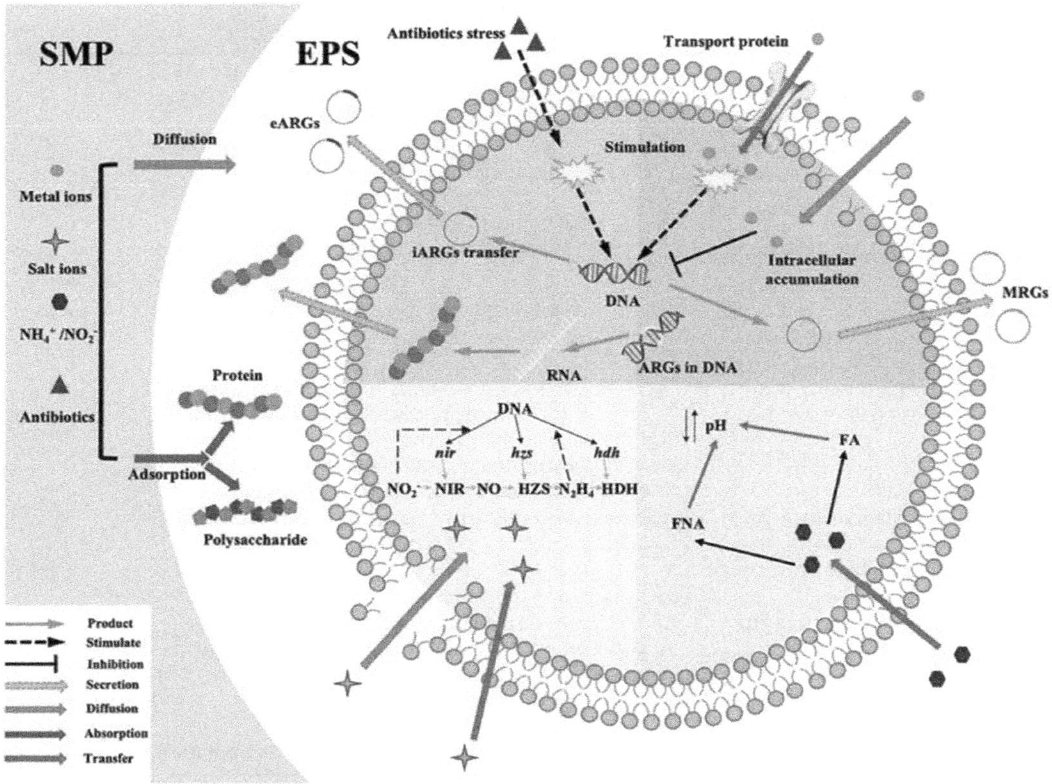

Figure 4.3 The mechanism of contaminant inhibition. eARGs: extracellular antibiotic resistance genes; iARGs: intracellular antibiotic resistance genes; MRGs: metal resistance genes; FA: free ammonia; FNA: free nitrous acid. Reproduced from ref. 26 with permission from Elsevier, Copyright 2022.

Table 4.1 Positive and negative impacts of different metals. Reproduced from ref. 26 with permission from Elsevier, Copyright 2022.

Metals	Inhibitory concentration	Positive effect	Negative effect
Zn^{2+}	6.90 mg L^{-1} (IC_{50})	Activate the enzyme active sites (<10 mg L^{-1})	Increase intracellular reactive oxygen species, resulting in cell leakage; reduce the content of EPS
Cu^{2+}	1.25 mg g^{-1} (Cu NPs) 74.30 mg L^{-1} (IC_{50})	Reduce electrostatic repulsion and bridging effect; promote the compaction and agglomeration of EPS colloid; involved in the synthesis of key enzymes; inhibit cell metabolism (<1 mg L^{-1})	Disrupt protein structure; catalyze the generation of hydroxyl free radicals; promote stress through the redox cycle, and destroy the cell membrane
Ag^{+}	11.52 mg L^{-1} (IC_{50})	No report	Inhibit the HDH enzyme activity and cytochrome *c* activity
Fe^{2+}	378.57 mg L^{-1}	Promote microbial activity and growth of essential trace elements; replace the substrate as a nutrient; function as growth promoter (1–5 mg L^{-1})	Inhibit the reaction rate of related enzymes
Cr^{6+}	10.91 mg L^{-1} (IC_{50})	Promote microbial growth (10 mg L^{-1})	Destroy the structure of DNA and proteins
Cd^{2+}	11.20 mg L^{-1} (IC_{50}) 5.43 mg L^{-1} (IC_{50})	Function as essential trace elements (2 mg L^{-1})	Inhibit the activity of enzyme; reduce the concentration of cytochrome *c*; disrupt protein structure
Ni^{2+}	48.60 mg L^{-1} (IC_{50}) 14.64 mg L^{-1} (IC_{50})	Promote the synthesis of dehydrogenation coenzyme (<1.74 mg L^{-1})	No report
Co^{2+}	8–10 mg L^{-1} (MAB)	Improve denitrification performance and increase the content of key enzymes in cells (0.06, 0.48 and 5.80 μg L^{-1})	Decrease microbial activity
Pb^{2+}	No effect	Function as essential trace elements (<40 mg L^{-1})	Reduce the concentration of cytochrome *c*;
Hg^{2+}	60.39 mg L^{-1} (IC_{50})	No report	Inhibit anammox bacteria activity; increase the abundance of heavy metal resistance genes
Mn^{2+}	4.83 mg L^{-1} (IC_{50})	Function as coenzyme factors; function as growth promoter	No report

4.4.2 Inhibition of Antibiotic Resistance Genes

Antibiotics stimulated the development of harmful bacteria at low doses but reduced the microbial activity at high concentrations. Antibiotics are frequently used in the healthcare business, aquaculture, and animal farming. Distinctive antibiotics, including chlortetracycline (20 ppm), tetracycline (1000 ppm), doxycycline (400 ppm), and acetaminophen (400 ppm), have different inhibitory concentrations. At minimal antibiotic concentrations, the amplitudes of AnAOB and denitrifying bacteria rose, and the reactors' efficiency of nitrogen removal was improved, according to prior research. Nevertheless, at high antibiotic doses, the amplitudes of AnAOB and denitrifying bacteria declined, whereas the population of nitrifying bacteria rose.[27] AnAOB produced additional EPS to preserve the cytoskeleton when antibiotic concentrations were low. However, EPS discharge was inhibited at more significant quantities, which might have altered the anammox process's efficacy considerably. The EPS content was reduced when oxytetracycline (OTC) 1000 ppm was administered through the anammox procedure. Once the concentration of tetracycline HCl reached 10 ppm, SAA was inhibited by 60%, and the NLR reduced from 5.81 ± 0.32 to 1.65 ± 0.24 kg N per m per day. The cytochrome *c* concentration dropped by half the minute the concentration of OTC was 2 ppm. 23–100 ppm OTC effectively suppressed the anammox action.[28] Antibiotics' inhibitory impact on microorganisms inhibits their activity and causes pollution. Antibiotic stress, for example, resulted in the buildup of ARGs and ARBs, which caused contamination. ARGs and ARBs endanger human health and have the potential to harm the environment. ARGs gather in bacteria and move across bacteria in various ways due to antibiotic selection pressure. The ecosystem and human health are both in danger due to their proliferation. Horizontal gene transfer (HGT) and vertical gene transfer (VGT) are both used to propagate ARGs in microbial communities.[29] HGT refers to the genetic propagation of ARGs during cell division. HGT by ARGs occurs *via* transformation, transduction, and conjugation. HGT uses mobile genetic elements (MGEs) like bacteriophages, plasmids, transposons, and integrons to spread ARGs. Plasmids are DNA molecules that are circular in shape passed from one species to another. Due to the apparent self-transport feature of plasmids, ARGs in combination with plasmids can transmit across species *via* the plasmids. ARGs may also cross genera and species by assembling on transposons. Novel transposon Tn 6433 variants have been discovered that hasten the spread of ARGs.[30] According to research, from infected bacteria, phages may collect ARGs and spread them to certain bacteria by transduction. According to the latest studies, biogas digestate ARGs are discharged into the ecosystem *via* aerosol dispersal.[31] Integrons may transport ARGs and disseminate in the environment. Extracellular ARGs (eARGs) and intracellular ARGs (iARGs) are the two common types of ARGs. Because external antibiotic resistance genes are created by secretion and cytolysis, iARGs were spread through conjugation and transduction, while eARGs were diffused by transformation. eARGs can be absorbed by bacteria, which means that eARGs can be transformed into beneficiary bacteria. Antibiotic resistance in microbial populations may be considerably increased through ARG transmission, presenting a hazard to ecological health.[1]

4.4.3 Inhibition of Emerging Pollutants

As the world becomes more urbanized, greater attention is being paid to the negative consequences of new contaminants, including nanoparticles, antibiotics, and microplastics. The above mentioned contaminants may be found in a variety of places in the environment, from aquatic environments to wastewater treatment facilities. AnAOB have also been shown to be harmful to emerging contaminants. Microplastics are described as plastic particles with a diameter of less than 5 mm that are created by shredding huge plastic pieces. Microplastics have an impact on the anammox process' nitrogen removal effectiveness and the structure of the microbial population at large levels.[32] In the ecosystem, there are two forms of microplastics: biodegradable and nonbiodegradable. Proteins and fibers, such as polyhydroxybutyrate, polylactic acid, and polybutenic acid, make up biodegradable microplastics. Microplastics influence the building integrity of EPS and prevent AnAOB from aggregating. The cell membrane in microorganisms serves a vital function in ensuring cell stabilization. Microplastics on the nanoscale adhere to cell membranes and quickly inhibit metabolic intermediates, critical enzymes, and finished outputs. Microplastics on the nanoscale kill cells and disrupt material transport pathways and DNA. Some other pollutants are incorporated by microplastics, such as polychlorinated biphenyls (PCBs), polycyclic aromatic hydrocarbons (PAHs), antibiotics, and metals due to their enormous specific surface area. The major source of damage from these contaminants is their desorption.[33] Heavy metals, antibiotics, and persistent organic pollutants are all carriers of microplastics. These pollutants impact AnAOB due to desorption from the microplastic exteriors. NPs are extensively employed in coatings, cosmetics, electronic goods, catalytic materials, and energy materials due to industrial and living necessities. Graphene oxides (GO), nanoscale zero-valent iron (nZVI), CuO NPs, TiO_2 NPs, Fe_3O_4 NPs, ZnO NPs, Ag NPs, and carbon nanotubes (CNTs) are among the most common NPs. Unfortunately, many of these NPs will inevitably be released into the surroundings and run into wastewater remediation facilities. Not all NPs affect the anammox process negatively.[34] Based on their concentration, NPs may have a favorable or detrimental impact on the anammox process. NPs improve the anammox process' nitrogen removal effectiveness and the diversification of the microbial population at low concentrations. Previous research has shown that nZVI helps anammox activity rebound after being inhibited by dissolved oxygen (DO).[35] Further research found that γ-Fe_2O_3

NPs and graphene nanocrystals (GNs) improved the anammox mechanism's nitrogen removal efficiency and boosted the functionality of hydrazine dehydrogenase (HDH) functionality by shortening the electron transport range and promoting the operation. γ-Fe_2O_3 NPs act as an electron transporter, shuttling electrons across various enzymes. The most critical phase in the process is electron transport, facilitated by GNs and NPs, hence substantially boosting the anammox process. NPs inhibit the anammox mechanism at high doses. To demonstrate this point, two instances are provided below. ZnO NPs are frequently employed in cosmetics, coatings, and pigments, and cause cell permeability and impair AnAOB survival by increasing intracellular reactive oxygen species production.[36] ZnO NPs decreased the performance of the anammox procedure, and the inhibitory impact amplified with the growing concentration of ZnO NPs. A wide range of elements are absorbed by ZnO NPs; for example, the bioavailability of Mo, As, Al, Pb, Hg, Cd, Cu, and Ni is altered in the solution by modification of their morphologies (Yamindago *et al.*, 2018). According to recent research, ZnO NPs have an IC_{50} of 84.7 ppm and a combined reversible inhibition dosage of 100 ppm. Some NPs hindered the anammox method performance. Metals like Cu^{2+} and Zn^{2+} were adsorbed by Ag NPs.[37] The deposited metals are then removed from the surface into the surroundings, increasing the metal concentration in the aqueous environment and increasing the anammox process's negative impacts. TiO_2 NPs show high absorptivity for Cd. GO increases oxidative stress by absorbing heavy metals and organic contaminants. Cu NPs engage with EPS *via* linking to groups comprising tryptophan or tyrosine *via* –OH and C–O–C bonds.[38] The promoting and inhibitory effects of different nanomaterials are shown in Table 4.2.

4.5 Challenges in the Anammox Process

4.5.1 The Out-selection of Nitrite-oxidizing Bacteria

The nitritation stage has a significant impact on the performance of anammox treatments. As a result, coordinating the various roles of bacterial communities is critical. The thorough out-selection of NOBs, including temperate zone and minimal nitrogen level settings, is thought to be among the most challenging jobs, particularly in the often variable NH_4^+–N loading that causes excess aeration. FA and FNA were implemented to monitor the nitritation operation in sidestream deammonification with high NH_4^+, but this strategic approach no longer assures fruitful mainstream mechanisms in a low level of nitrogen. Earlier, it was observed that the NO_2^- oxidation by NOB was hindered at FA concentrations of 0.1–1.0 ppm, whereas the AOB-mediated NH_4^+ oxidation was blocked at 10–150 ppm.[14] Furthermore, at FA concentrations of 1.0–10 ppm, selective suppression of NOB through NO_2^- build-up might occur. At an FA concentration of 25 ppm, Balmelle *et al.*[39] found that the

Table 4.2 Promoting and inhibitory effects of different nanomaterials. Reproduced from ref. 26 with permission from Elsevier, Copyright 2022.

Nanomaterials	Application	Inhibitory effect	Promoting effect
nZVI	Repair soil and groundwater pollution	Destroy cell structure and affect cell activity (100 $mg\,L^{-1}$)	Alleviate the inhibition of dissolved oxygen on anammox bacteria (<1.70 $mg\,L^{-1}$)
CuO NPs	Catalytic materials, semiconductor materials	Inhibit the removal performance of nitrogen, COD and phosphorus (1.4 $mg\,L^{-1}$)	No report
GO	Energy materials, catalytic material, biosensor material	Destroy the cell membrane structure, and affect the metabolic activity of microorganisms (<5 $mg\,L^{-1}$)	Promote the denitrification performance of anammox bacteria; improve the activity of anammox bacteria; agglomerate structure of anammox granules; improve the activity of HDH enzyme; promote the electron transfer process (100 $\mu g\,L^{-1}$)
TiO_2 NPs	Catalysts, cosmetics, coatings	Inhibit the removal performance of COD, nitrogen and phosphorus; inhibit the activity of enzymes (50 $mg\,L^{-1}$)	Agglomerate structure of anammox granules
Fe_3O_4 NPs	Used as conducting material	Inhibit the nitrogen removal performance of the anammox process	Enhance the activity of anammox bacteria and promote the electron transfer process (<200 $mg\,L^{-1}$)
ZnO NPs	Catalysts, cosmetics, coatings	Inhibit the denitrification performance of COD, nitrogen and phosphorus (11.6 $mg\,g\,VSS^{-1}$)	No report

AOB performance was decreased by 40%, but Kim *et al.*[40] found that the NOB activity was lowered by more than 50% at an FA concentration of 0.7 ppm. To achieve a steady NO_2^- deposition rate of 84 percent, an FA concentration of 10.6 mg NH_3 per L was advised. Nitrifying organisms were suppressed by FNA concentrations of 0.22–2.8 ppm due to the existence of nitrous acid. Under anoxic circumstances, Wang *et al.* (2014)[41] found that FNA was much more hazardous to NOB than AOB. Though pH regulation is an option, due to the high chemical cost, this is not a monetarily viable method of nitritation. Generally, adjusting the operating temperature >25 °C to grow AOB substantially quicker than NOB is a good technique for managing nitritation. On the other hand, this method is not realistic for full-scale applications. Most traditional nitritation engineering solutions have relied on managing the DO of bioreactors, which entails functioning DO at 0.5–0.9 ppm and then decreasing DO to 0.06 ppm to identify the NO_2^- or NH_4^+ lessening slopes at the lower DO level. Nevertheless, apparatus dependability, for instance, an air blower, is crucial for the process's steady functioning.[42]

4.5.2 High Ratio of Organic Carbon to Nitrogen

Traditional household influent wastewater has an extremely high C/N ratio (ranging from 10–12), making it inappropriate for mainstream anammox/nitritation procedures. The ideal proportion for a partial nitritation/anammox (PN/A) therapy is generally between 2 and 5, or even less than 0.5. The presence of large organic substance levels in the wastewater inlet inhibits the development of anammox bacteria, but pinpointing the specific causes for this singularity is difficult. "Metabolic pathway conversion inhibition" and "inhibition and out-competition" have been proposed as reasons for organic carbon's inhibitory influence on the anammox process.[43] Initially, Chamchoi *et al.* (2008)[44] and Güven *et al.* (2005)[45] found that heterotrophic bacteria may grow more rapidly than anammox bacteria at high COD concentrations since they competed for space and resources, implying that anammox bacteria performance reduces due to territorial and nutrient constraints. Chen *et al.* (2016) also showed that adding glucose to an anammox reactor stimulates denitrifying bacteria's growth, while anammox bacteria's growth is unaltered by supplementation of glucose.[46] Furthermore, the anammox effect can be considerably lower throughout the process of nitrogen removal, suggesting that anammox bacteria digested organic carbons instead of NO_2^- and NH_4^+ as feedstock. Anammox bacteria showed altered metabolic capabilities for "metabolic pathway conversion inhibition" when acetate was added as a substrate throughout reactor operation. A C/N ratio of 2.7–2.9 lowered both the anammox and AOB activities. But when the C/N ratio was raised from 1.1 to 1.5, the mechanism of nitrogen degradation amplified due to a rise in the population of anammox bacteria. However, their growth rate was reduced once the C/N ratio was 2.0. These findings suggest that the restricted generation of NO_2^- is to be blamed for these events.[47] Furthermore, at a 0.8 C/N ratio, AOB may be flushed out of the reactor. The effectiveness of the anammox/nitritation method for nitrogen removal declined from 79 to 52% as the wastewater C/N ratio grew from 0.5 to 0.75. The efficacy of the anammox techniques with regard to nitrogen removal efficiency (NRE) for processing mainline and sidestream wastewater at various C/N ratios can be differentiated depending on the working statistics obtained in lab-scale units. Anammox processes have significant NRE in mainline and sidestream wastewater treatment (71–96%) at low C/N ratios (0.0–1.0). Once the C/N ratio exceeded 1.0, the NRE in sidestream anammox procedures steadily fell (58.0–85.4%), but it dramatically decreased (4–77.3%) in mainstream processes. The NREs in mainstream anammox structures are often less reliable than those in sidestream ones.[2] The performance of the anammox process for treating mainstream-like wastewater under different organic carbon to nitrogen (C/N) ratios is shown in Table 4.3.

4.5.3 Retention of Anammox Bacteria

Due to the low growth rates of autotrophs, a biomass retaining strategy must be effectively maintained. This ensures a healthy microbial population for promising start-up and system stabilization of anammox operations. Due to the absence of a pristine culture, the precise doubling time of anammox bacteria is unknown. The doubling period of anammox bacteria has been observed to vary from 10 to 12 days in different investigations. The estimated doubling periods may extend to 25–30 days based on the variety of species and growth circumstances, or even at the full-scale/pilot in the temperate zone.[48] Moreover, since the quantity of NH_4^+ in urban wastewater is so low, starting and maintaining anammox activity is difficult. As a result, a prolonged SRT should be used to effectively manage the maintenance of anammox biomass in the mainline anammox/nitritation method, particularly at lower than 20 °C. *Anammoxoglobus*, *Brocadia*, and *Kuenenia* species have been detected in both the natural habitat and presently operating wastewater treatment plants.[3] *Brocadia* is among the most widely found species in mainline and sidestream wastewater treatment anammox systems (along with MBBRs SBRs, UASB reactors, concurrent nitrification, denitrification reactors, anammox reactors, and other membrane bioreactors). Notably, despite the fact that the Brocadia is the earlier anammox species to be discovered, the *Kuenenia* species is the maximum and most widely researched in biochemistry, physiology, cell structure, and relative genomics study.[49] Recent studies have experimentally confirmed the involvement of major proteins in the genome of *Kuenenia stuttgartiensis*, such as hydrazine dehydrogenase, hydrazine synthase, hydroxylamine dehydrogenase, and an S-layer protein with an encrypted putative anammox pathway.[50,51] The experimental verification of critical step hydrazine generated from

Table 4.3 Performance of the anammox process for treating mainstream-like wastewater under different organic carbon to nitrogen (C/N) ratios. Reproduced from ref. 2 with permission from Elsevier, Copyright 2021.

Reactor type	C/N (mg COD mg^{-1} N)	Inf. TN (mg N L^{-1})	NRE (%)	Problems
SBR	0.6–2.0	22.0 ± 5.9	70–40	COD increased, the diffusion limitation impacted AnAOB more than NOB, reduced on both the AOB and AnAOB activities
SBR	2–2.5	59.3	73.5–77.3	AnAOB activity stopped increasing as increasing amounts of nitrogen were removed
	1.1–2.0	59.3	30.8–73.5	AnAOB activity and abundance increased
NRBC	0.5–0.75	100–200	79–52	Decrease in nitrogen removal efficiency
MBBR	3.2	21.8 ± 5.2	73–91	No interruption because nitrogen load was less
UASB	1.1	45.9	88.4	Lower carbon resulted in high removal rate
H-MBBR	3.2	21.8 ± 5.2	63	Less stable in removing nitrogen due to increase in nitrate produced

HNO_2 was the most intriguing characteristic of the anammox species; this implies that the *Kuenenia* species may have unique nitrogen metabolic pathways. Furthermore, *Kuenenia* species that dwell in freshwater habitats are supplemented by anammox-operated wastewater treatment plants. For instance, Akaboci *et al.* (2018) found that *Kuenenia* species predominate in a grainy one-step fractional anammox/nitritation SRT at a temperate zone (15 °C), indicating that these species might be necessary for anammox operations.[52] Likewise, Gu *et al.* (2018) found that *Kuenenia* species were enhanced in the biofilm of the MBBR system throughout the procedure of A-B processes, which include an anaerobic fixed-bed reactor and an SBR.[53] Also, Yang *et al.* (2018) found that the Kuenenia species dominated over the Brocadia species when partial nitritation/anammox methods were shifted from the side flow to the mainline.[54] The challenges for mainstream anammox process application are shown in Figure 4.4.

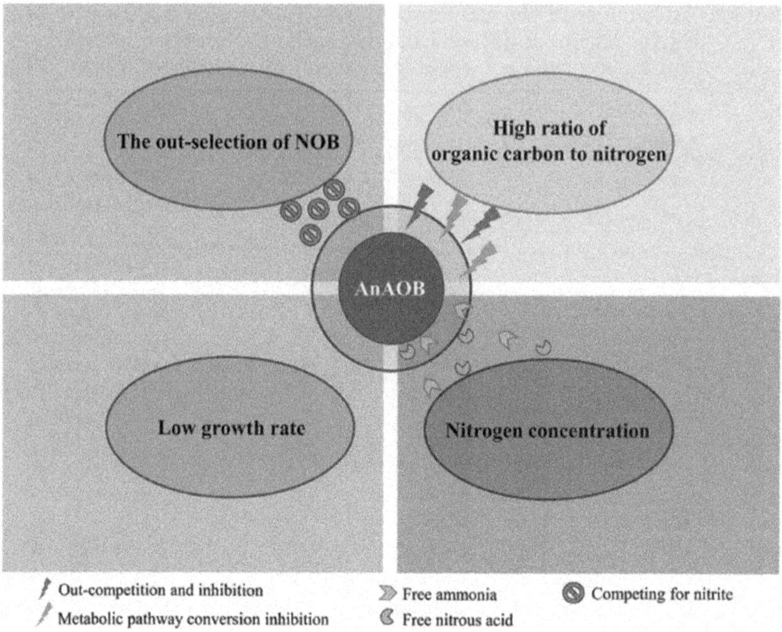

Figure 4.4 Challenges for mainstream anammox process applications. Reproduced from ref. 2 with permission from Elsevier, Copyright 2021.

4.6 Opportunities for the Application of Anammox Processes

4.6.1 Performance Improvement of Low-strength Wastewater

The parameters of minor wastewater treatment have a significant impact on partial nitritation and the anammox procedures. The nitritation phase is regarded as one of the critical processes since it efficiently picks and develops the AOB, suppresses the NOB, and achieves a 50 percent oxidation of NH_4^+ to NO_2^-. Remediation of various forms of wastewater is quite challenging, particularly in cold areas. Many autotrophic systems to remove nitrogen reportedly worked at temperatures more than 25 °C with the concentration of nitrogen flowing in exceeding 100 mg N L^{-1}. Even though the proportion of NOB and AOB in the sludge remains essentially consistent, the extremely competitive NOB may benefit more from a drop in temperature due to their increased productivity and formation of NO_3^-.[55] When the working temperature is dropped below 25 °C, the SAA is dramatically reduced. Furthermore, in low NH_4^+ concentrations (often less than 50 mg L^{-1}), the essentially minuscule FA might hinder anammox activities. As a result, investigating low nitrogen level and low temperature in wastewater offers fresh possibilities to use anammox procedures for designing energy-efficient WWTPs. A growing curiosity in anammox study in low nitrogen concentration and cold environments is visible. *Brocadia* bacteria (*e.g.*, *Brocadia fulgida* and *Brocadia sinica*) are often found in systems that run at temperatures between 5 and 6 degrees Celsius. *Kuenenia stuttgartiensis* was similarly detected in a constant bioreactor arrangement, which functioned at low temperature (>20 °C). A modified anammox mush at a low temperature has a greater effective anammox rate, according to Lotti *et al.* (2015).[56] The PN/A process obtained an NRR of 30 mg N per L per day in a standard nutrient removal range structure when run at 15 °C on pre-treated municipal wastewater *via* a 21 ± 5 mg NH_4^+–N per L nitrogen flow in. Limiting DO to 2.5 mg O_2 per L to regulate NOB might result in effective nitritation inhibition at both 20 and 15 °C, resulting in adequate NRRs of 0.40 and 0.44 g N Total per L per day and respectively optimum NREs of 73 and 86 percent. Furthermore, an assessed growth rate of 0.017 per day was reported under the same circumstances at 15 °C, indicating that the methods may be used in mainline WWTPs.[57] During a typical IFAS deammonification procedure, 10 percent of the principal outflow wastewater was directed straight to the deammonification reactor first before biological phosphorus and COD removal steps to increase the inlet flow C/N ratio from 2.3 to 3.1 g (sCOD) per gNH_4^+–N. By this process, 72 percent of total nitrogen degradation was achieved when the temperature dropped from 20 to 12 degrees Celsius, and anammox breakdown handled nitrogen removal of around 53 percent, and no detrimental effects on anammox activity were noticed due to the carbon content increase in the inlet wastewater.[58] The NOB and AOB were concentrated especially in the flocculate of a fused PN/A MBBR setup with 23 mg L^{-1} NH_4^+, whereas in the biofilm, anammox bacteria were concentrated. The growth rates of NOB and AOB fell, and NOB was washed out of the flocs, but the tiny NOB component persisted in the biofilm, hindered *via* lower DO content (from 1.2 to 0.17 mg L^{-1}). The anammox colony has been proposed to be triggered in low temperature and nitrogen circumstances, and it turns out to be a viable setting for use in mainline WWTPs.[59]

4.6.2 Management of High C/N Ratio for Maximizing C Energy Recovery

COD is typically converted to renewable energy using anaerobic digestion techniques. The largest CH_4 produced from biodegradable COD is 0.35 N m^3 CH_4 per kg COD, which may be utilized to power thermal plants and motors and generate revenue for the wastewater-generating business as well as WWTPs. Water low in strength and temperature might stymie this direct use, and recovering chemical energy is a major difficulty.[60] The main principle is exemplified in the A–B procedures, which were intended to optimize the collecting capability of carbon components out of wastewater prior to passing across autotrophic metabolisms as proposed by Li *et al.* (2018).[61] The "A" process involves the anaerobic process, CEPT (chemically enhanced primary treatment), HRAS (high rate activated sludge), IFAS (integrated fixed-film activated sludge), EBPR (enhanced biological phosphorus removal), bioelectrochemical arrangement, biodegradation of COD, and methanization. Since organic mass is not required for anammox denitrification, it is converted into methane for power generation, resulting in a 47 percent growth in methane yield and a 50% decline in energy consumption for air supply compared to standard biotic nitrogen degradation strategies.[62] The "B" process is involved in the treatment of the residual nutrients (nitrogen) that pass across the autotrophic metabolic track. For anaerobic treatment, a UASB reactor has been used to pre-treat the primary wastewater at 20 °C, and afterward, the wastewater was screened employing a fabric filter (20 μm pore size) to retrieve NH_4^+ concentrations and soluble COD (sCOD) in an inflow range of 35–50 mg N per day and 44–88 mg (sCOD) per L, respectively. HRAS procedures have a strong capability for producing high food-to-microorganism (F/M) ratios containing COD at a reasonably low SRT and short HRT. The high-rate aerobic system HRAS procedures condensed the inflow sCOD, particulate, and colloidal matter as sludge, and they can subsequently be used to recover energy through anaerobic breakdown.[63] Wett *et al.* (2013) used an in-depth HRAS to extract COD prior to its use in mainline aerobic reactors, and anammox bacteria and AOB derived from a sidestream DEMON reactor.[64] By choosing a high-density sludge fraction, a hydro cyclone classifier was employed to hold anammox particles, whereas alternating aeration was used to suppress NOB activities. As a result, when full-scale HRAS processes were run for 1 HRT hour and 0.3 days, they removed about 90% of the water's Biochemical Oxygen Demand (BOD). They then used a one-step anammox/nitritation setup at 19 °C for ten months to remove the remaining NRE. The EBPR process was simultaneously used to remove organic phosphorus and carbon from household wastewater. A setup with double SBRs treated practical domestic sewage by combining improved phosphorus removal using a biological technique with nitration and anammox.[65] When there were alternating periods of anaerobic–aerobic conditions, the first SBR removed phosphorus with 2–3 days of SRT to avoid nitrification. The nitrate/anammox SBR was then in charge of getting rid of the residual NH_4^+, with a maximum NRR of 0.12 kg N per m^3 per day. In the CEPT method, chemicals like metal salts and polymers are added to primary clarifiers to make them coagulate and clump together.[66] However, the CEPT might not be able to remove sCOD, which could impact the anammox/nitritation activity. Before designing a CEPT as a pre-treatment, predefined things should be looked at very carefully. The bioelectrochemical framework can be used as pre-treatment before making energy or removing nitrogen through current generation. A microbial fuel cell system was used to handle the suspended particles over 90% and the total COD of 75%. The residual nitrogen in the anode effluent was therefore processed in the cathode using nitritation–anammox.[67]

4.6.3 Commercialized Processes for Mainstream Treatment

The real-world use of anammox procedures is restricted through its exploration progress due to a shortage of pristine cultures, stringent operating requirements, and the poor anammox bacteria growth rate. The use of anammox in nitrogen removal operations, on the other hand, would substitute the traditional denitrification stage, reducing the nitrification aeration cost by half and allowing for a decrease in operating expenses. The first full-sized granulated anammox device with a feeding rate of 750 kg N per day was installed for sludge handling in 2002 at the Rotterdam MWWTP, Netherlands. It emblematically marked the start of the industrial use of anammox procedures.[68] Furthermore, the single-step fractional anammox/nitritation reactor ($4\,m^3$) utilized to support the solid waste to A-stage of the WWTP was handled independently in winter (13.4 ± 1.1 °C) and summer (23.2 ± 1.3 °C), and it demonstrated total NRR values of 0.097 ± 0.016 and 0.223 ± 0.029 kg N m^3 per day.[69] According to certain experiments, heterotrophic biomass was easily excreted from the system. In contrast, anammox bacteria developed, keeping standard WWTP conditions in mind, and particles were efficiently enriched and kept in the setup. These findings suggest that the projected granular sludge reactor has a strong chance of being effective in removing autotrophic nitrogen from WWTPs' main streams.[70] Several effective routes for full-scale use and enrichment of anammox in actual treatment supported this achievement of mainstream wastewater treatment with effective full anammox implementation on high nitrogen content wastewater. An exemplary life-size improved nitrogen removal for mainstream wastewater by *in situ* amplification of anammox bacteria was effectively implemented in the MWWTP at Xi'an in Shanxi province, China, with a size of 250 000 m^3 per day^1. Furthermore, its anaerobic–anoxic–oxic (HRT 2.0, 2.6 and 6.4 h) procedure was improved by including 25% moving carriers to the anoxic zone, outfitting submerged mixers to assure proper contact among the carriers and wastewater, and configuring impeding screens in the exit to avoid the suspended carrier's loss.[71] As the inlet COD/TN ratio fluctuated (from 1.2 to 7.9), effective removal of nitrogen was carried out with a 7.6 ppm mean discharge concentration of total inorganic nitrogen (TIN) after roughly 2 years of constant operation. Furthermore, metagenome sequencing study revealed that the comparative ampleness of anoxic-carrier biofilm anammox bacteria was substantially greater than that documented in ordinary WWTP flocculent sludge. The Strass WWTP (Austria) attained steady mainline anammox techniques by amplifying anammox biomass and AOB from sideline SBR, anammox bacteria preservation by hydrocyclone, and aeration control to constrained NOB. These were implied by high nitrite concentrations instead of nitrate in the outflow and measured by aerobic activity tests (75% inhibition).[72] The Glarnerland WWTP demonstrated the practicality of the deammonification idea by obtaining higher than 90% ammonia removal efficiency (max load of 200 kg NH_4^+–N), quite similar to the Strass WWTP.[73] In the Marselisborg WWTP (Aarhus, Denmark), experiments were done to see whether anammox might supplement the usual denitrification/nitrification methods in a low-temperature region. Using ^{15}N-labeling studies at various temperatures (10, 20, and 30 °C), it was discovered that anammox caused around 1% of NRE after 3 years of operation.[74] The lack of adaptability might ascribe this discovery to the temperate zone and the minimal number of anammox bacteria in the reactors. Because of that, the anammox procedures were precluded by the dominating nitrification. Similarly, the ^{15}N-labeling rate demonstrated the evident influence of anammox performance on temperature (7.7 ± 0.6 and 72.9 ± 7.2 µmol g VSS per h at 10 °C and 30 °C, respectively). One more experiment that utilized two-step mainline deammonification procedures not using bioaugmentation to polish nitrogen was undertaken in the WWTP of Chesapeake Elizabeth, USA. Partly nitrified discharged wastewater was run as the initial stage activated sludge system by adjusting the NH_4^+ to NO_2^- and nitrate (NO_x–N) ratio, and subsequently anammox MBBR as the next stage. The system used transitory aeration with an elevated DO of 1.0 mg L^{-1} and a brief SRT of roughly 0.1 d at 20 °C. An 8 : 11 ratio of COD/NH_4 also hindered the nitrogen removal effectiveness of anammox MBBR polishing. In Singapore, the Changi Water Reclamation Plant (200 000 m^3 per day^1) used intermittent aeration to oxidize around 72% NH_4^+ to NO_2^-, which in the anoxic zone, was removed by anammox methods.[75] The ANITA™ Mox procedures effectively treated mainline outflow with 65 mg L^{-1} sCOD content *via* IFAS arrangement (DO – 0.2–0.8 ppm) following the HRAS method. For mainstream applications, the facile and robust separation technique of IFAS ANITA™ Mox keeps anammox biomass in the reactor. Pilot data indicate that treatment effectiveness suppresses NOB performance, whereas the N removal rate increases at 16–18 °C up to 0.2–0.3 kg N m^3 per day^1. In recent years, Aqualia (AquELAN® with granulated biomass), DC water (DEMON® with granulated biomass), DEMON GmbH, Paques (ANNAMOX® with granulated biomass), and Veolia (ANITA® Mox with IFAS and MBBR) have all worked hard to produce viable anammox products.[76] This eco-friendly and cost-effective technology is projected to become a common wastewater treatment approach in the nearish term. Figure 4.5 represents the efficient approaches for anammox enrichment and full-scale application in practical treatment.

4.7 Future Prospects of Anammox Application

New findings and research have dramatically enhanced the knowledge of mainline anammox mechanisms during the last decade. However, implementing these procedures on a large scale demands technological ways to solve urgent issues. First and foremost, these techniques should address the issue of using anammox procedures on low nitrogen concentration effluent. In municipal wastewater with less nitrogen inputs, the partial anammox/nitrification procedure may be regarded to function in stable circumstances.

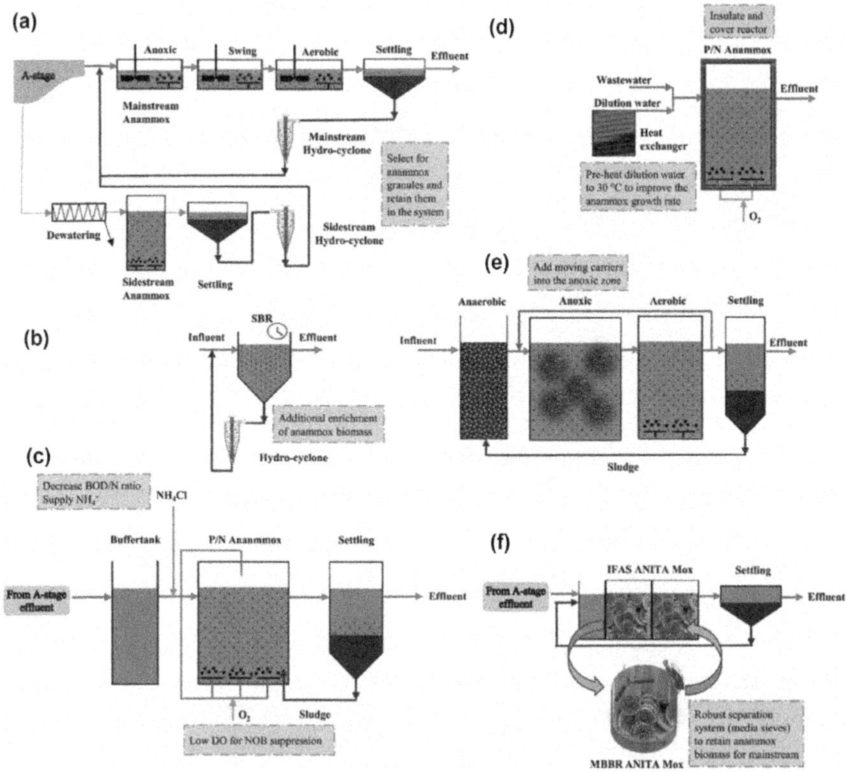

Figure 4.5 Efficient approaches for anammox enrichment and full-scale application in practical treatment. (a) Hydrocyclones in combination with the mainstream and sidestream system; (b) hydrocyclones for enriched biomass in SBR; (c) chemical supply for the improvement of processes; (d) dilution of water and heat exchange for the improvement of growth rate; (e) moving carriers for the enhancement of efficiency; and (f) robust separation for retaining biomass. Reproduced from ref. 2 with permission from Elsevier, Copyright 2021.

Second, removing NOB from the treating mechanism is a difficult task. Because it is challenging to regain activities quickly while boosting anammox performance throughout the shift from anoxic to aerobic, especially at cold temperatures, alternate aeration is a helpful method used in mainline anammox procedures for NOB impediment. Besides, FNA and FA have lately been recommended for NOB screening in mainline operations, owing to their extra substantial biocidal impact on NOB than AnAOB and AOB. Third, an anammox reactor should maintain a minimum organic carbon content since organic substances destructively influence the reaction. The A-B method was created to enhance the entrapment capability of carbon bases from inlet wastewater to control high C/N ratios prior to the anammox reaction. The A-B process's anaerobic reactor aids in reducing organic carbon content and aids in the creation of methane by anaerobic metabolism. Biogas containing methane can be used to generate electricity and heat, and that can be used to operate wastewater treatment facilities. The addition of NH_4Cl may help to enhance low-intensity influents, while also lowering the C/N ratio. However, the operating costs of obtaining stabilization should be thoroughly considered when using this chemical technique. Finally, the most challenging task is to preserve AnAOB's gradual expansion in mainstream anammox processes. Multiple strategies that have proved successful in recent times may be used to assure the proper maintenance of AnAOB. SBR, MBBR, membrane bioreactors, and full-size reactor systems effectively stored large concentrations of AnAOB, as per lab experiments, and they offered operating flexibility. Physical setups including mechanical vibrating screens, hydro-cyclones, and sieving equipment might be chosen as a cost-effective solution to retain AnAOB in mainstream anammox operations. Companies may generate marketable goods and obtain sludges loaded with AnAOB for real-world uses to mainline anammox procedures using the described methodologies.

4.8 Conclusion

Anammox is a cost-effective and efficient biological wastewater remediation method. Nonetheless, the requirements for AnAOB survivability are demanding, and AnAOB is quite susceptible to toxins in the atmosphere. The effects of ammonia, nitrite, and organic contents on the anammox process have been thoroughly explored and evaluated. Metals, antibiotics, chemical compounds, NPs, and microplastics are the primary contaminants that impact the anammox process. The techniques that determine whether such contaminants impede or enhance the anammox process are currently unknown. These contaminants may have a beneficial or harmful impact on the anammox procedure. Through the introduction of transitional goods or suitable solutes, some of the detrimental consequences of impurities on the anammox procedure might be mitigated. Metagenomics, metatranscriptomics, proteomics, and glycomics will be used to investigate AnAOB's response to inhibition. In addition, this study highlights the problems and potential of mainline anammox procedures by assessing their investigational findings and related potential as published in current articles. The difficulties of high organic C/N ratio, NOB out-selection, and retaining stunted AnAOB have been recognized as three crucial problems for applying the anammox method to mainstream wastewater. To address these issues, feasible and effective techniques have been offered. To solve the primary issues, anammox, and fractional nitritation, organic carbon removal before anammox, aeration control, and the physical parting of AnAOB have been proposed. Outlining the fundamental problems and routes for overcoming them is essential to explore the potential research paths for the growth of mainline anammox procedures. There are just a few occasions when mainline anammox procedures have been used on a big scale. Nevertheless, as discussed in this chapter, it is envisaged that mainstream anammox procedures will become widespread in the nearish term if new practices are done with answers to the obstacles that impede their implementation.

References

1. D. Q. Huang, J. J. Fu, Z. Y. Li, N. S. Fan and R. C. Jin, *Sci. Total Environ.*, 2021, DOI: 10.1016/j.scitotenv.2021.150009.
2. H. P. Trinh, S. H. Lee, G. Jeong, H. Yoon and H. D. Park, *J. Environ. Chem. Eng.*, 2021, DOI: 10.1016/j.jece.2021.105583.
3. A. Gonzalez-Martinez, B. Muñoz-Palazon, A. Rodriguez-Sanchez and J. Gonzalez-Lopez, *FEMS Microbiol. Lett.*, 2018, DOI: 10.1093/femsle/fny031.
4. X. Li, S. Klaus, C. Bott and Z. He, *Water Environ. Res.*, 2018, DOI: 10.2175/106143017X15131012153112.
5. Y. Guo, Y. Chen, E. Webeck and Y. Y. Li, *Bioresour. Technol.*, 2020, DOI: 10.1016/j.biortech.2019.122560.
6. Z. Kong, L. Li, Y. Xue, M. Yang and Y. Y. Li, *J. Cleaner Prod.*, 2019, DOI: 10.1016/j.jclepro.2019.05.233.
7. Z. Wenjie, D. Yuhan, W. Dunqiu, C. Heng and J. Yue, *Global NEST J.*, 2016, DOI: 10.30955/gnj.002048.
8. J. Zhang, Q. Zhang, Y. Miao, Y. Sun, J. Chen and Y. Peng, *Bioresour. Technol.*, 2018, DOI: 10.1016/j.biortech.2018.09.110.
9. C. J. Tang, P. Zheng, C. H. Wang, Q. Mahmood, J. Q. Zhang, X. G. Chen, L. Zhang and J. W. Chen, *Water Res.*, 2011, DOI: 10.1016/j.watres.2010.08.018.
10. H. Chen, C. Ma, G. F. Yang, H. Z. Wang, Z. M. Yu and R. C. Jin, *Bioresour. Technol.*, 2014, DOI: 10.1016/j.biortech.2014.06.063.
11. Y. Guo, Y. Zhao, T. Zhu, J. Li, Y. Feng, H. Zhao and S. Liu, *Water Res.*, 2018, DOI: 10.1016/j.watres.2018.06.052.
12. S. Lackner, E. M. Gilbert, S. E. Vlaeminck, A. Joss, H. Horn and M. C. M. van Loosdrecht, *Water Res.*, 2014, DOI: 10.1016/j.watres.2014.02.032.
13. S. Wang, Y. Liu, Q. Niu, J. Ji, T. Hojo and Y. Y. Li, *Bioresour. Technol.*, 2017, DOI: 10.1016/j.biortech.2017.03.164.
14. A. C. Anthonisen, R. C. Loehr, T. B. S. Prakasam and E. G. Srinath, *J. – Water Pollut. Control Fed.*, 1976, DOI: 10.1205/j.watpoll.1976.02.054.
15. Y. Zhou, A. Oehmen, M. Lim, V. Vadivelu and W. J. Ng, *Water Res.*, 2011, DOI: 10.1016/j.watres.2011.06.025.
16. M. C. De Prá, A. Kunz, M. Bortoli, T. Perondi and A. Chini, *J. Chem. Technol. Biotechnol.*, 2012, DOI: 10.1002/jctb.3803.
17. B. Molinuevo, M. C. García, D. Karakashev and I. Angelidaki, *Bioresour. Technol.*, 2009, DOI: 10.1016/j.biortech.2008.10.038.
18. T. Kindaichi, T. Awata, Y. Mugimoto, R. M. L. D. Rathnayake, S. Kasahara and H. Satoh, *Chemosphere*, 2016, DOI: 10.1016/j.chemosphere.2016.06.018.
19. S. W. How, A. S. M. Chua, G. C. Ngoh, T. Nittami and T. P. Curtis, *Sci. Total Environ.*, 2019, DOI: 10.1016/j.scitotenv.2019.07.332.

20. Q. Zhang, S. E. Vlaeminck, C. DeBarbadillo, C. Su, A. Al-Omari, B. Wett, T. Pümpel, A. Shaw, K. Chandran, S. Murthy and H. D. Clippeleir, *Water Res.*, 2018, DOI: 10.1016/j.watres.2018.06.037.
21. X. Zhang, Z. Chen, Y. Ma, J. Zhao, T. Chen, H. Fu and H. Zhai, *Bioresour. Technol.*, 2018, DOI: 10.1016/j.biortech.2018.04.055.
22. C. Yu, X. Tang, L. S. Li, X. L. Chai, R. Xiao, D. Wu, C. J. Tang and L. Y. Chai, *Bioresour. Technol.*, 2019, DOI: 10.1016/j.biortech.2019.03.081.
23. X. Ma, Y. Yan, W. Wang, J. Guo and Y. Wang, *Chemosphere*, 2020, DOI: 10.1016/j.chemosphere.2019.125682.
24. Q. Q. Zhang, X. M. Ji, G. M. Tian and R. C. Jin, *Bioresour. Technol.*, 2021, DOI: 10.1016/j.biortech.2020.124106.
25. Q. Q. Zhang, H. Qian, P. Y. Li, J. Q. Zhao, Y. Q. Sun and R. C. Jin, *Bioresour. Technol.*, 2021, DOI: 10.1016/j.biortech.2021.124945.
26. D. Q. Huang, J. J. Fu, Z. Y. Li, N. S. Fan and R. C. Jin, *Sci. Total Environ.*, 2022, DOI: 10.1016/j.scitotenv.2021.150009.
27. Y. Meng, B. Sheng and F. Meng, *Sci. Total Environ.*, 2019, DOI: 10.1016/j.scitotenv.2019.02.389.
28. Q. Q. Zhang, H. Chen, J. H. Liu, B. E. Yang, W. M. Ni and R. C. Jin, *Bioresour. Technol.*, 2014, DOI: 10.1016/j.biortech.2013.11.053.
29. S. Aydin, B. Ince and O. Ince, *Water Res.*, 2015, DOI: 10.1016/j.watres.2015.07.007.
30. Y. Shi, Y. Shi, Z. Tian, Z. Tian, M. R. Gillings, Y. Zhang, Y. Zhang, H. Zhang, J. Huyan, J. Huyan, M. Yang and M. Yang, *Environ. Sci. Technol.*, 2020, DOI: 10.1021/acs.est.0c01272.
31. Y. Zhang, Y. Zheng, Z. Zhu, Y. Chen and H. Dong, *Sci. Total Environ.*, 2021, DOI: 10.1016/j.scitotenv.2020.144108.
32. L. Tang, C. Su, Y. Chen, Y. Xian, X. Hui, Z. Ye, M. Chen, F. Zhu and H. Zhong, *J. Hazard. Mater.*, 2021, DOI: 10.1016/j.jhazmat.2020.123337.
33. Z. Zhang and Y. Chen, *Chem. Eng. J.*, 2020, DOI: 10.1016/j.cej.2019.122955.
34. A. Elreedy, S. Ismail, M. Ali, S. Q. Ni, M. Fujii and M. Elsamadony, *J. Environ. Manage.*, 2021, DOI: 10.1016/j.jenvman.2020.111495.
35. Y. Yan, Y. Wang, W. Wang, S. Zhou, J. Wang and J. Guo, *Water Res.*, 2019, DOI: 10.1016/j.watres.2019.01.029.
36. Z. Z. Zhang, Y. F. Cheng, L. Z. J. Xu, Y. H. Bai, J. J. Xu, Z. J. Shi, Q. Q. Zhang and R. C. Jin, *Sci. Total Environ.*, 2018, DOI: 10.1016/j.scitotenv.2017.12.016.
37. W. Liu, Z. Zeng, A. Chen, G. Zeng, R. Xiao, Z. Guo, F. Yi, Z. Huang, K. He and L. Hu, *J. Environ. Chem. Eng.*, 2018, DOI: 10.1016/j.jece.2018.06.032.
38. Z. Z. Zhang, J. J. Xu, Z. J. Shi, Y. F. Cheng, Z. Q. Ji, R. Deng and R. C. Jin, *Bioresour. Technol.*, 2017, DOI: 10.1016/j.biortech.2017.03.135.
39. B. Balmelle, K. M. Nguyen, B. Capdeville, J. C. Cornier and A. Deguin, *Water Sci. Technol.*, 1992, DOI: 10.2166/wst.1992.0543.
40. D. J. Kim, D. I. Lee and J. Keller, *Bioresour. Technol.*, 2006, DOI: 10.1016/j.biortech.2005.03.032.

41. Q. Wang, L. Ye, G. Jiang, S. Hu and Z. Yuan, *Water Res.*, 2014, DOI: 10.1016/j.watres.2014.02.029.
42. R. Wen, Y. Jin and W. Zhang, *Int. J. Environ. Res. Public Health*, 2020, DOI: 10.3390/ijerph17031090.
43. R. C. Jin, G. F. Yang, J. J. Yu and P. Zheng, *Chem. Eng. J.*, 2012, DOI: 10.1016/j.cej.2012.05.014.
44. N. Chamchoi, S. Nitisoravut and J. E. Schmidt, *Bioresour. Technol.*, 2008, DOI: 10.1016/j.biortech.2007.08.029.
45. D. Guven, A. Dapena, B. Kartal, M. C. Schmid, B. Maas, K. Van De Pas-schoonen, S. Sozen, R. Mendez, H. J. M. O. Den Camp, M. S. M. Jetten, M. Strous and I. Schmidt, *Physiol. Biotechnol.*, 2005, DOI: 10.1128/AEM.71.2.1066-1071.2005.
46. C. Chen, F. Sun, H. Zhang, J. Wang, Y. Shen and X. Liang, *Bioresour. Technol.*, 2016, DOI: 10.1016/j.biortech.2016.05.115.
47. M. Han, H. De Clippeleir, A. Al-Omari, B. Wett, S. E. Vlaeminck, C. Bott and S. Murthy, *Water Sci. Technol.*, 2016, DOI: 10.2166/wst.2016.202.
48. T. L. G. Hendrickx, Y. Wang, C. Kampman, G. Zeeman, H. Temmink and C. J. N. Buisman, *Water Res.*, 2012, DOI: 10.1016/j.watres.2012.01.037.
49. S. H. Peeters and L. van Niftrik, *Curr. Opin. Chem. Biol.*, 2019, DOI: 10.1016/j.cbpa.2018.09.022.
50. A. Dietl, C. Ferousi, W. J. Maalcke, A. Menzel, S. De Vries, J. T. Keltjens, M. S. M. Jetten, B. Kartal and T. R. M. Barends, *Nature*, 2015, DOI: 10.1038/nature15517.
51. W. J. Maalcke, J. Reimann, S. De Vries, J. N. Butt, A. Dietl, N. Kip, U. Mersdorf, T. R. M. Barends, M. S. M. Jetten, J. T. Keltjens and B. Kartal, *J. Biol. Chem.*, 2016, DOI: 10.1074/jbc.M116.735530.
52. T. R. V. Akaboci, F. Gich, M. Ruscalleda, M. D. Balaguer and J. Colprim, *Chem. Eng. J.*, 2018, DOI: 10.1016/j.cej.2018.05.115.
53. J. Gu, Q. Yang and Y. Liu, *Water Res.*, 2018, DOI: 10.1016/j.watres.2018.02.051.
54. Y. Yang, L. Zhang, J. Cheng, S. Zhang, X. Li and Y. Peng, *Bioresour. Technol.*, 2018, DOI: 10.1016/j.biortech.2017.12.079.
55. R. C. Jin, G. F. Yang, Q. Q. Zhang, C. Ma, J. J. Yu and B. S. Xing, *Water Res.*, 2013, DOI: 10.1016/j.watres.2012.12.018.
56. T. Lotti, R. Kleerebezem and M. C. M. van Loosdrecht, *Biotechnol. Bioeng.*, 2015, DOI: 10.1002/bit.25333.
57. H. Siegrist, D. Salzgeber, J. Eugster and A. Joss, *Water Sci. Technol.*, 2008, DOI: 10.2166/wst.2008.048.
58. P. Roots, A. F. Rosenthal, Q. Yuan, Y. Wang, F. Yang, J. A. Kozak, H. Zhang and G. F. Wells, *Environ. Sci.: Water Res. Technol.*, 2020, DOI: 10.1039/D0EW00652A.
59. M. Laureni, D. G. Weissbrodt, K. Villez, O. Robin, N. de Jonge, A. Rosenthal, G. Wells, J. L. Nielsen, E. Morgenroth and A. Joss, *Water Res.*, 2019, DOI: 10.1016/j.watres.2018.12.051.
60. R. P. Taylor, C. L. Jones and R. K. Laubscher, *J. Water Process Eng.*, 2021, DOI: 10.1016/j.jwpe.2020.101797.

61. X. Li, S. Klaus, C. Bott and Z. He, *Water Environ. Res.*, 2018, DOI: 10.2175/106143017X15131012153112.
62. P. L. McCarty, *Environ. Sci. Technol.*, 2018, DOI: 10.1021/acs.est.7b05832.
63. J. Jimenez, M. Miller, C. Bott, S. Murthy, H. De Clippeleir and B. Wett, *Water Res.*, 2015, DOI: 10.1016/j.watres.2015.07.032.
64. B. Wett, A. Omari, S. M. Podmirseg, M. Han, O. Akintayo, M. Gómez Brandón, S. Murthy, C. Bott, M. Hell, I. Takács, G. Nyhuis and M. O'Shaughnessy, *Water Sci. Technol.*, 2013, DOI: 10.2166/wst.2013.150.
65. Y. Yang, J. Gao and Y. Zhang, *Optik*, 2017, DOI: 10.1016/j.ijleo.2017.03.027.
66. V. Diamantis, W. Verstraete, A. Eftaxias, B. Bundervoet, V. Siegfried, P. Melidis and A. Aivasidis, *Water Sci. Technol.*, 2013, DOI: 10.2166/wst.2013.639.
67. Z. Ge and Z. He, *Environ. Sci.: Water Res. Technol.*, 2016, DOI: 10.1039/C6EW00020G.
68. W. R. L. van der Star, W. R. Abma, D. Blommers, J. W. Mulder, T. Tokutomi, M. Strous, C. Picioreanu and M. C. M. van Loosdrecht, *Water Res.*, 2007, DOI: 10.1016/j.watres.2007.03.044.
69. M. Hoekstra, S. P. Geilvoet, T. L. G. Hendrickx, C. S. van Erp Taalman Kip, R. Kleerebezem and M. C. M. van Loosdrecht, *Environ. Technol.*, 2019, DOI: 10.1080/09593330.2018.1470204.
70. T. Lotti, R. Kleerebezem, Z. Hu, B. Kartal, M. K. De Kreuk, C. Van Erp Taalman Kip, J. Kruit, T. L. G. Hendrickx and M. C. M. Van Loosdrecht, *Environ. Technol.*, 2015, DOI: 10.1080/09593330.2014.982722.
71. J. Li, Y. Peng, L. Zhang, J. Liu, X. Wang, R. Gao, L. Pang and Y. Zhou, *Water Res.*, 2019, DOI: 10.1016/j.watres.2019.05.070.
72. B. Wett, S. M. Podmirseg, M. Gómez-Brandón, M. Hell, G. Nyhuis, C. Bott and S. Murthy, *Water Environ. Res.*, 2015, DOI: 10.2175/106143015X14362865227319.
73. B. Wett, G. Nyhuis, I. Takács and S. Murthy, *Proc. – Water Environ. Fed*, 2012, DOI: 10.2587/10305Y143629191324.
74. A. Kamp, L. D. M. Ottosen, N. B. Thøgersen, N. P. Revsbech, B. Thamdrup and M. H. Andersen, *Water Sci. Technol.*, 2019, DOI: 10.2166/wst.2019.141.
75. C. Yeshi, K. B. Hong, M. C. M. van Loosdrecht, G. T. Daigger, P. H. Yi, Y. L. Wah, C. S. Chye and Y. A. Ghani, *Water Sci. Technol.*, 2016, DOI: 10.2166/wst.2016.116.
76. T. Liu, S. Hu and J. Guo, *Crit. Rev. Biotechnol.*, 2019, DOI: 10.1080/07388551.2019.1598333.

CHAPTER 5

Ammonia Oxidizing Bacteria in Wastewater Treatment

PRANJAL P. DAS,*[a] MUKESH SHARMA[b] AND MIHIR K. PURKAIT[a]

[a] Department of Chemical Engineering, Indian Institute of Technology Guwahati, Assam, India; [b] CORAL, Indian Institute of Technology Kharagpur, West Bengal, India
*Email: pranj176107027@iitg.ac.in

5.1 Introduction

In the last few years, the presence of uncontrolled trace organic compounds (TrOCs) or emerging micro-pollutants in water bodies has raised significant international concerns due to their bioaccumulative, poor biodegradable, persistent and hazardous nature. Pharmaceutically active compounds, endocrine disruptive chemicals, steroidal hormones, artificial sweeteners, personal care products, industrial chemicals, insecticides, and their metabolites are among the several organic categories (natural or anthropogenic) that exist in aquatic systems. The quantity and type of these substances in both natural and artificial environments are closely linked to scattered sources or the population's socio-economic level. Hospital or pharmaceutical industry discharges, municipal (domestic) effluent, livestock and aquaculture, direct discharge of agricultural runoff, laboratory wastewater, and urban and highway runoff are common sources of such substances entering vital water bodies.[1] Figure 5.1 represents the sources and possible fate of TrOCs in aquatic ecosystems. In reality, present wastewater treatment facilities (WWTFs) are primarily designed to degrade macropollutants and are based on aerobic and anaerobic systems.

Ammonia Oxidizing Bacteria: Applications in Industrial Wastewater Treatment
Edited by Maulin P. Shah

Published by the Royal Society of Chemistry, www.rsc.org

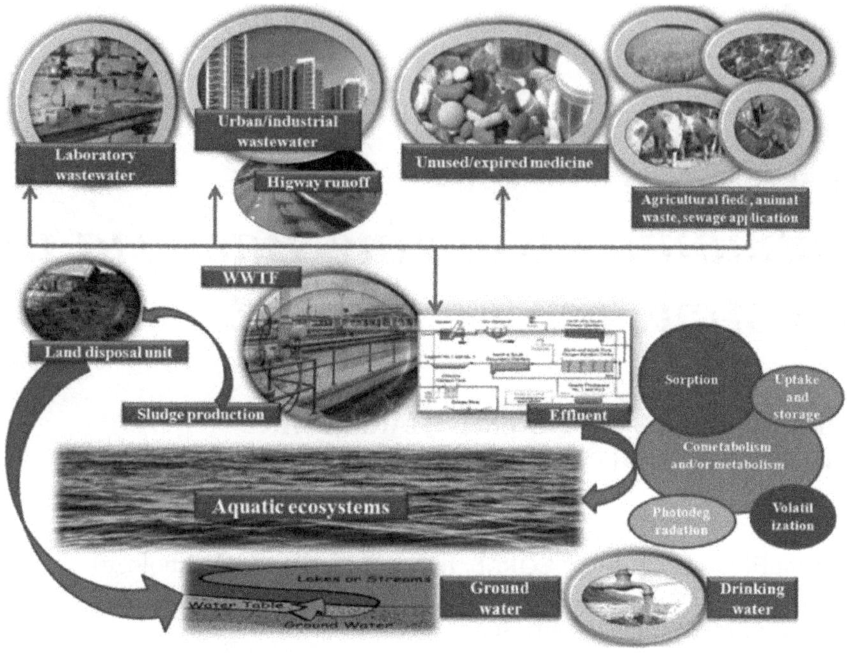

Figure 5.1 Sources and possible fate of TrOCs in aquatic ecosystems. Reproduced from ref. 6 with permission from Elsevier, Copyright 2019.

Therefore, most TrOCs, especially persistent and hazardous TrOCs, are poorly removed by traditional treatment plants. In such situations, partly degraded substances in WWTFs are likely to be identified in several environmental compartments across the world, *viz.* drinking water, soil/sediment, groundwater, and surface and sewage water, where TrOCs can create adverse consequences even at low concentrations ($ng\,L^{-1}$ or less).[2] Given the known environmental concerns, measures for reducing the development of such compounds or degrading the TrOCs in both natural and artificial systems may be necessary. Biodegradation, photo-transformation, volatilization and adsorption are some of the methods that have been found to remove TrOCs. Because of its low cost and the ability to completely degrade TrOCs, biodegradation remains one of the best solutions. Degradation of TrOCs in biological techniques is highly variable and has recently been described to be firmly linked with significant ammonium oxidizing activities obtained throughout the biological wastewater treatment process. The reason may be attributed to the diverse metabolic activity of nitrifying bacteria, especially ammonia-oxidizing bacteria (AOB). Through cometabolic biodegradation, the non-specific enzyme ammonia monooxygenase (AMO) can digest a wide spectrum of aromatic and aliphatic compounds.[3] The biodegradation of TrOCs *via* AOB was detected in nitrifying activated sludge (NAS) when significant ammonia oxidizing activities were obtained, thereby resulting in the degradation of a wide range of TrOCs. It has also been reported that heterotrophic microorganisms possess the ability to degrade TrOCs either alone or in combination with AOB, predominantly *via* metabolic pathways. As a result, when the systems are fabricated and operated with nitrification capability rather than non-nitrifying systems, the abatement of certain compounds is highly possible. To date, only a few review studies on the removal process, behaviour, occurrence, and fate of TrOCs in artificial and natural systems have been published in international journals. In this context, Tran *et al.* (2013) evaluated the metabolism and cometabolism routes for micropollutant biodegradation.[4] Also, pharmaceutical compounds were extensively removed by AOB throughout the biological treatment operations, according to Xu *et al.* (2016).[5] Nevertheless, research devoted to interpreting the role of AOB to increase the biodegradation of TrOCs in NAS systems, their analysis and parameters effecting the degradation process has fallen far behind. As a result, it is necessary to examine and discuss the current developments and understanding on the degradation of TrOCs by AOB.

The present chapter elaborately reviews the state of the art development of AOB and their ecophysiology in both artificial and natural environments. The key parameters governing the removal of TrOCs and the various degradation routes have been discussed and summarized in detail. Furthermore, the biodegradation mechanism of AOB along with the biodegraded products have been demonstrated. TrOCs have been characterized as compounds that are predominantly employed for diagnostic, preventative and therapeutic purposes and are commonly found and/or evaluated in both wastewater effluents and surface water. Thus, the role of AOB in the treatment of different wastewater has also been taken into consideration. Also, different potential strategies for achieving sustained contaminant removal have been examined. As there exist more stringent discharge limits for TrOCs, a critical evaluation of the specific functions of AOB in TrOC degradation will be advantageous in enhancing our understanding of the complexities involved in its biotransformation.

5.2 Ammonia-oxidizing Bacteria: An Overview

Reduction in the nitrogen content as in ammonia is utilized as a coherent source of energy, CO_2 as a suitable carbon source, and molecular O_2 as an acceptor of electrons by bacteria capable of oxidizing ammonia. These chemolithotrophic AOB are mostly Beta and Gamma proteobacteria, like Nitrosococcus (Gamma), Nitrosomonas (Beta), and Nitrosospira (Beta). From the investigation of amoA and 16S rRNA gene sequences obtained from the cultures and the environmental clones, Nitrosomonas/Nitrosospira species seem to be governing both the natural and artificial habitat. As a result, AOB from Proteobacteria in microbial ecology research, the beta subclass, have been employed as a model organism.[7] However, because the composition of the complete AOB community has yet to be determined, molecular probes are unlikely to find the full AOB group. Furthermore, few AOB exhibit anaerobic metabolism. Planctomycetales species employ nitrite as an electron acceptor rather than oxygen, resulting in N_2 formation. In any N-containing habitat along with a significant anoxic zone, the anammox process tends to be much more active.[8]

Anammox bacteria are members of the phylum Planctomycetes exhibiting slow metabolism. The cytoplasm of anammox bacterial cells is separated into three membranes, according to their physiology. The cytoplasmic thin membrane is the outermost layer, which surrounds the inner side of the cell wall. Furthermore, the other membrane, which is called the intracytoplasmic membrane, isolates the riboplasm from the paryphoplasm, whereas anammoxosome is the third thin membrane separating the anammox machinery from the riboplasm. Furthermore, the cell diameters of different anammox-related genera vary greatly. Anammox cells from the genera *Candidatus Kuenenia stuttgartiensis* and *Candidatus Brocadia fulgida* have an average diameter of 880 nm and anammoxosomes occupy 61% of the cell content. The reported average diameter of *Candidatus Scalindua* spp. is around 950 nm. Furthermore, as reported, the largest cell diameter (1100 nm) belongs to the genus *Candidatus Anammoxoglobus propionicus*.[9] Anammox species have a diverse metabolic system. Even though it was previously thought that anammox bacteria used nitrite as an electron acceptor to oxidize ammonia, new research reveals that anammox bacteria may employ an alternate electron acceptor instead of nitrite. Organic acids (propionate) compete with the ammonium for electron acceptor nitrite. *Anammoxoglobus propionicus* was proposed as the name of the novel anammox after more research. It may be concluded from the recently studied anammox strain that the anammox method has a wide variety of uses for removing nitrogen and organic acids from wastewater. Anammox can dissimulate nitrate reduction to ammonium DNRA, according to a recent study.[10] Anammox bacteria's metabolism is influenced by their surroundings and is dictated by pollutants, including the ammonium, sulphate, propionic acid, urea, and others, demonstrating the metabolic diversity of anammox species. Kartal *et al.* (2011) investigated the exact mechanism of anammox metabolism in the anammox species *Kuenenia stuttgartiensis*.[11] Hydrazine is made using nitric oxide rather than hydroxylamine as a precursor. Nitrite reduction to NO has been attributed to the cd1 nitrite/nitric oxide oxidoreductase (NirS). Even though the genes associated with the generation of hydroxylamine are unidentified, it is hypothesized that the Hao-like protein is essential for the transformation of nitrite to hydroxylamine. Even though inclusion of the Nrf enzyme in anammox bacteria was confirmed, its relevance is still unclear. The Nrf enzyme detoxifies hydroxylamine and NO_2^- to ammonium or oxidizes the organic compound employing nitrite as an electron acceptor for synthesizing ammonium in anammox metabolism.[12]

5.3 Ammonia-oxidizing Bacteria (AOB) and Nitrification

5.3.1 Primary Enzymes of AOB

AMO, or hydroxylamine oxidoreductase, is the most essential enzyme for transforming NH_4 to hydroxylamine (NH_2OH), which gets further transformed to NO_2^- (HAO). HAO is a one-of-a-kind enzyme that catalyses the oxidation of the hydroxylamine to NO_2 during the second phase of nitrification, utilizing water as a terminal electron acceptor and extramolecular oxygen as a cofactor. HAO is a soluble enzyme that exists in the bacteria's periplasm but is attached to the cytoplasmic membrane. It was revealed that AMO carries Cu ions.[13] The active site of AMO contains Fe ions as well. The AMO enzyme is made up of 3 protein subunits that are encoded by the genes amoB, amoC, and amoA. Even though all three amo genes could be studied for molecular investigation, only a subset of amoA operates as an operational gene of the AOA and AOB, the enzymes that catalyse the initial reaction in AO, and has been extensively applied as a biomarker gene to examine the abundance and diversity of AOB. The amoA gene is the most important and has a well-established database, as it generates a protein with the AMO active site and gives energy for AOB production. It is followed by amoB and amoC.[14] Other genetic variants in ammonia oxidation (AmoB and AmoC) should be investigated further. By cometabolic biodegradation, AMO can eradicate a range of TrOCs, namely methane, hydrocarbons, halogenated, alkenes, and aromatic hydrocarbons, and other micropollutants. Furthermore, AMO enzymes produced from both AOB and AOA contribute to organic pollutants' cometabolic biodegradation.[15] Future investigations must concentrate on how archaeal and bacterial AMO participate for the cometabolic biological decomposition of the pollutants.

5.3.2 Ammonia-oxidizing Bacteria Detection and Identification

Since the early 1990s, the application of advanced molecular technologies has eliminated many of the constraints related with identifying and isolating AOB and other groups of microorganisms, along with providing information on biodegradation and the microbiological ecology within any environmental sample. AOB have traditionally been isolated using various culture methods, which takes time because to the poor biomass yield rates of AOB. AO isolates have been accessible since the 1900s, but research on their diversity in wild nitrifying populations did not begin until culture-independent molecular methods for microbial environmental surveys became available.

Denaturing real-time quantitative PCR (qPCR), gradient gel electrophoresis (DGGE), and the production and sequencing of 16S rRNA gene libraries are instances of culture-independent molecular technologies that have assisted in the investigation of AOB, and other communities in a variety of habitats.[6] Head *et al.* (1993) for the very first time employed PCR-amplified 16S rRNA gene sequences extracted from pure cultures of the representative strains to detect AOB.[16] The 16S rRNA gene was first discovered in water samples and used to analyse AOB populations. The identification and classification of AOB was achieved through the development of group-specific primers and probes for the 16S rRNA gene. Since then, the 16S rRNA gene is widely recognized to study AOB phylogeny. The AmoA genes are widely employed as a basic functional marker in the study of AO species. The amoA and 16S rRNA gene phylogenies, despite their agreement, show several discrepancies. To measure autotrophic AO within a freshwater lake, wastewater treatment plant, and enrichment culture, Rotthauwe *et al.* (1997) used a polymerase chain reaction (PCR)-based methodology.[17] In a variety of circumstances, DGGE and real-time PCR assays have been employed to assess AOB and other community diversity. The quantity and composition of AOB communities have been studied using DGGE of PCR amplicons of the 16S rDNA gene. However, because of amplification of 16S rRNA genes from much similar non-nitrifying bacteria, the 16S rDNA gene technique for the ecological studies of AOB may result in an overprediction of both the diversity and number of AOB. Furthermore, Aakra *et al.*, (2001) proposed implementing combined 16S rDNA and amoA techniques to investigate wild AOB populations since the target genes should be the most accurate.[18] The molecular approaches for detecting AOB in the environment include FISH-microautoradiography (FISH-MAR), fluorescence *in situ* hybridization (FISH), terminal restriction fragment length polymorphism (T-RFLP), isotope array, live/dead BacLight (LD), and quinone profiling.

5.4 Role of Ammonia Oxidizing Bacteria in Wastewater Treatment

The function of the AOB in the biodegradation of the bulk of TrOCs in artificial environments has gotten a lot of attention. Along with the essential AOB's cometabolic biodegradation, nitrifying WWTPs typically outperformed non-nitrifying WWTPs in degrading many TrOCs. This increased clearance appears to be linked to AOB activity, which could be co-metabolizing TrOCs *via* AMO. Bisphenol A, iohexol, naproxen, terbutryn, and irgarol, for example, have been demonstrated to deteriorate much faster in the nitrified system than in the non-nitrified system in several investigations.[19] In numerous systems, including NAS and MBRs, nitrification has been demonstrated to drive the biodegradation of medicines and hormones.[20] De Gusseme *et al.* (2009) observed >94% EE2 biodegradation in both the synthetic sewage and WWTF effluent by the process of nitrifying enrichment culture of MBR, whereas no significant degradation by heterotrophic organisms isolated from this culture was reported.[21] Park *et al.* (2017) found that when AOB activity was not inhibited, the trimethoprim elimination efficiency was higher (70%) than when AOB activity was suppressed (28%).[22] Ibuprofen, trimethoprim, and naproxen clearance rates improved as the NH_4 loading rate increased, boosting AOB's biological activity and allowing it to create more AMO. Similarly, Xu *et al.* (2016) revealed that sludge-based systems with substantial nitrification seemed to have a 90% removal efficiency for the pharmaceuticals assessed than those without.[23] Limiting the enzyme amo along with the ATU in AOB-enhanced AS was used to investigate the effect of AOB in suppressing TrOCs in both pure cultures and activated sludge. Pure cultures of nitrifying bacteria can just destroy Bisphenol A in the absence of ATU. The rate of biodegradation of TrOCs has been measured using AOB. Sulfamethoxazole was well decomposed (86%) by the enriched AOB during aerobic batch experiments, resulting in higher nitrification rates. Additionally, a positive linear correlation among nitrifying activity and pollution abatement rate has also been established for a wide range of pollutants (*e.g.*, atenolol, roxithromycin, ibuprofen, 17-ethinylestradiol, and erythromycin).[24] *Nitrosomonas europaea* was shown by Shi *et al.* (2004) to be capable of removing steroidal estrogens from sewage.[25] A variety of TrOCs were explored to degrade better in the unavailability of allylthiourea, triclosan, and artificial sweeteners than in the availability of allylthiourea ($>$40% removal). Ammonia-oxidizing and heterotrophic bacteria have been discovered to cause bisphenol A breakdown. Similarly, the presence of ATU had no effect on ketoprofen or naproxen breakdown, and it's suspected that both AOBs and heterotrophic bacteria are participating in destroying them. Similarly, the nitrification inhibitor ATU reduced lincomycin clearance.[26] In the absence of AOB inhibition, the removal of 2-quinoxalinecarboxylic acid and mefenamic acid was substantially enhanced: 85% of mefenamic acid and 86% of 2QCA were eliminated, contrary to just 22% and 29% in the presence of AOB inhibition. In both instances (with and without AOB inhibition), theophylline and caffeine were eliminated, demonstrating that microbial activity in MBR treatment efficiently cleared these compounds independent of the presence of AOB. To sum up, biological processes that are administered with ammonium eliminate TrOCs more effectively than those not treated with ammonium. The biodegradation of TrOCs by autotrophic and heterotrophic organisms changes depending upon the types of TrOCs.[6]

5.5 Ammonia-oxidizing Bacteria Biodegradation

5.5.1 Mechanism of Biodegradation

The primary technique for removing the majority of TrOCs has been found as biodegradation. For example, larger mixed liquid suspended solid concentrations produced in MBR by increased SRT can improve TrOC

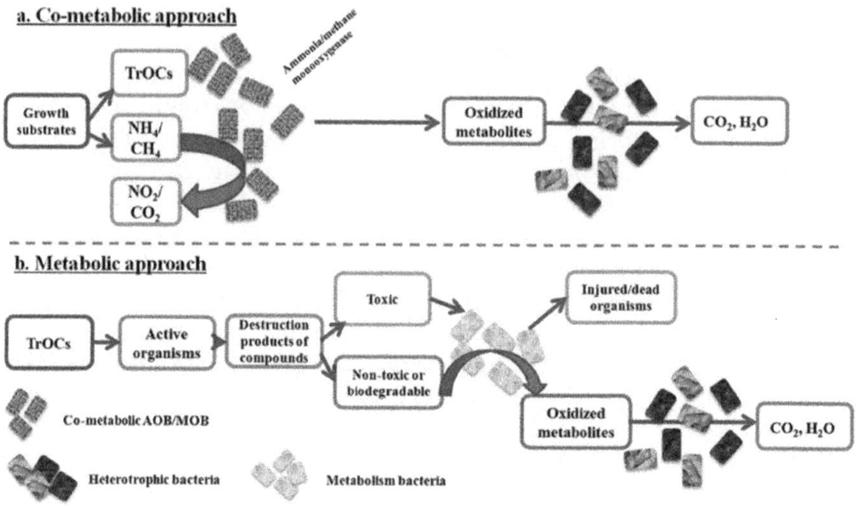

Figure 5.2 A schematic illustration of the metabolic and cometabolic approaches of trace organic chemical (TrOC) biodegradation. Reproduced from ref. 6 with permission from Elsevier, Copyright 2019.

biodegradation. TrOC biodegradation can occur *via* two primary processes (*e.g.*, metabolism and cometabolism). Cometabolic and metabolic response stages may be closely linked and interchangeable. Both cometabolic and metabolic strategies were recognized to neutralize numerous xenobiotics. Moreover, autotrophic bacteria cometabolize TrOCs in both natural as well as artificial environments, but heterotrophs remove TrOCs *viz.* cometabolism and/or metabolism pathways.[27] A schematic diagram of the metabolic and cometabolic pathways of TrOC biodegradation is shown in Figure 5.2.

5.5.1.1 *Metabolic Pathways*

TrOCs are the source of energy and/or carbon for organisms in the AS system, and they provide the necessary oxidation/reduction enzymes. So as to be digested by the organisms, TrOCs shouldn't be toxic for the development of the microbes and must be available in a sufficient concentration to facilitate the preservation of the biomass along with the generation of the much required cofactors and enzymes for the breakdown mechanism. Through metabolic breakdown, most TrOCs are rarely employed as the only energy and carbon source. To date, investigations on the metabolic biodegradation of TrOCs as the primary source of carbon have only been partially examined.[28] Quintana *et al.* (2005) investigated the organic breakdown of five acidic medicines in aerobic circumstances using sludge as an inoculum.[29] From the investigation it was revealed that primarily ketoprofen indicated metabolic biodegradation, whilst bezafibrate, ibuprofen, and naproxen only displayed oxidative biodegradation. Iasur-Kruh *et al.* (2011)

explored microbes that are capable of using several TrOCs, for instance ketoprofen and estradiol, as their only sources of energy and carbon.[30] These researchers also offered evidence of metabolic pathways for some of these chemicals' breakdown. However, no attempt has been made so far to give a comprehensive analysis of autotrophic organism metabolism. According to the most highly cited study on heterotrophic organisms' metabolic clearance of TrOCs, most TrOC elimination follows a cometabolic route with NH_4 being the primary source of energy/substrate. A significant metabolic limitation for bacteria is the incapacity of degradative enzymes to catabolize certain TrOCs.[31]

5.5.1.2 *Cometabolism Pathways*

Cometabolism is the process of transforming a non-growth substrate, TrOC, into a progress substrate for broadening the breadth and the range of TrOC removal. The nature and quantity of the growth substrates (such as toluene and phenol) as well as the concentration of NH_4 can all have an impact on the cometabolic strategy. As a result, the elimination of TrOCs is inextricably linked to the removal of growth substrates (NH_4).[32] The cometabolic process *via* AOB outperformed the metabolic process in terms of TrOC removal. Micropollutants are typically identified in engineering systems at concentrations ranging from nanograms to micrograms per litre, with some of them harmful to species. In this situation, TrOCs are an ineffective medium for growth. Henceforth, cometabolic biodegradation is assumed as the most prevalent organic decomposition method, which may be advantageous for most TrOCs because of their small concentrations. AMO cometabolism was recognized as a key stage in the TrOC elimination process.[33] According to Hooper *et al.* (1997), the *Nitrosomonas europaea* AMO enzyme can cometabolize more than 35 halogenated chemicals, along with NH_4 serving as a microbial growth substrate and source of energy.[34] Wahman *et al.* (2006) illustrated that *N. europaea* may cometabolize trihalomethanes in both artificial and natural environments.[35] Cometabolic biodegradation was conducted for a wide range of compounds in various environments, both aerobically and anaerobically. The aerobic cometabolic biodegradation of TrOCs is increasingly getting prominence, and is considered to be more preferable than removal *via* the anaerobic cometabolic pathway. In designed systems, these chemicals are degraded by both heterotrophs and autotrophs, such as AOB.[36]

5.5.2 Products and Pathways of TrOC Biodegradation

TrOCs are physiologically removed to variable degrees, resulting in mineralization or incomplete removal. The metabolites are linked to the operating circumstances and can be as persistent and harmful as, if not more so, than their parent chemicals. AOB cometabolically degrade TrOCs in nitrifying cultures, according to numerous studies. When compared to degradation without ammonium, AOB appear to drive cometabolic degradation and result in the synthesis of a variety of biotransformation products and pathways, indicating that cometabolism may have an impact on the development of TrOC degradation products.[37] Quintana *et al.* (2005) discovered that bezafibrate, ibuprofen, and naproxen were only biotransformed cometabolically in laboratory biodegradation tests utilizing activated sludge.[29] Iopromide could be converted into a variety of compounds, both with and without nitrification. The availability of heterotrophs and AOB was also observed to be associated with the biotransformation of several TrOCs. In enriched nitrifying cultures, Khunjar *et al.* (2011) confirmed this collaborative participation in the creation of 17-ethinylestradiol biotransformation products.[38] Trimethoprim was oxidized to hydroxytrimethoprim during the biotransformation of activated sludge, while trimethoprim remained active throughout the treatment. The biodegradation materials of penciclovir proved to be the consequence of enzyme processes including hydroxyl group oxidation and oxidation followed by acetate cleavage. Several medications' biodegradation products and processes are also influenced by their architecture. Xu *et al.* (2017) advised breakdown products and pathways of atenolol based on the biotransformation components reported, for developing an improved nitrifying culture.[39] The biodegradation of atenolol was linked to nitrifying organism activity in the availability of NH_4, illustrating that AOB cometabolize in the enhanced nitrifying sludge. There were four metabolites or compounds produced at the time of biodegradation of the atenolol along with oxidation

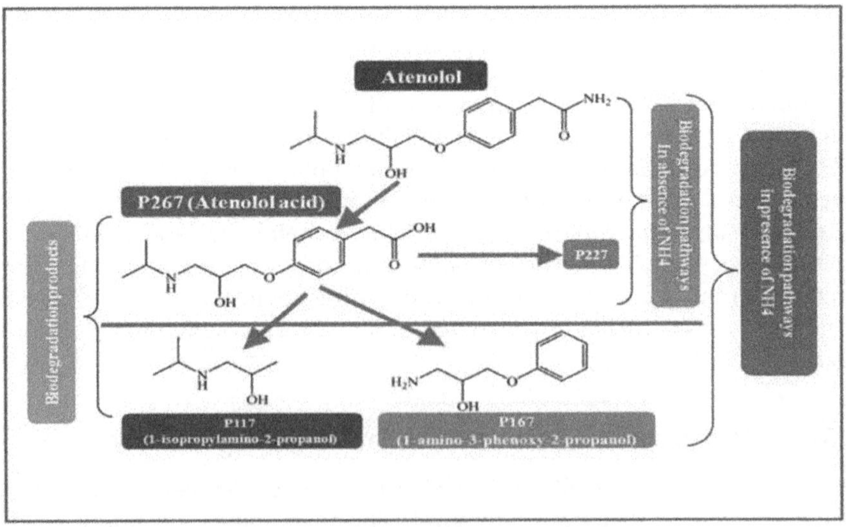

Figure 5.3 Pathways and products proposed for biodegradation of atenolol by an enriched nitrifying sludge. Reproduced from ref. 6 with permission from Elsevier, Copyright 2019.

of ammonia. The pathways and products proposed for biodegradation of atenolol by an enriched nitrifying sludge are shown in Figure 5.3.

5.6 Strategies for Sustained Pollutant Removal

5.6.1 Microbial Diversity and Cometabolism Strategy Development

In order to enhance the removal of TrOCs from artificial and natural and systems, complex microbial consortia made up of autotrophic and heterotrophic microorganisms could be employed. AOB and heterotrophs may be able to work cooperatively to break down TrOCs, according to several research studies. Meanwhile, studies have identified the functions of autotrophic and heterotrophic organisms, as well as their enzymes, in cometabolism and micropollutant metabolism.[40] A wide range of TrOCs were harmed by the AMO. The quantity and qualities of relevant degradative enzymes generated by these species, on the other hand, perform an essential function of the elimination of hazardous toxic chemicals. Although heterotrophic bacteria can successfully eliminate a range of TrOCs *viz.* monooxygenases and/or dioxygenases, enhanced TrOC removal by autotrophic AO *viz.* AMO has been well illustrated.[41] In a similar investigation, Khunjar *et al.* (2011) found that these two organisms remove micropollutants at different rates, with AOB outperforming heterotrophic species.[38] In a nitrifying process, such as WWTF, or in a natural process containing AOB and heterotrophic species, both organisms may breakdown the TrOCs. Overall, using microbial diversity in AS systems can improve TrOC cometabolic activity.

Cometabolic and metabolic response stages can be inextricably linked and interchangeable. Cometabolism is suspected as the principal process for the majority of the micropollutant removal because TrOC concentrations seems to be extremely low to act as growth substrates. Previous research hasn't been able to discriminate between the metabolic and cometabolic components of TrOC degradation.[42] Other bacteria than heterotrophic and autotrophic bacteria can eliminate these substances through cometabolic elimination. However, only heterotrophic bacteria can be seen using a metabolic method. Using organisms' cometabolic approach, a new strategy for degrading TrOCs might be created. Since the cometabolic degradation may occur in anaerobic/aerobic systems in biological wastewater treatment processes, the cometabolic activities of autotrophic and heterotrophic bacteria may aid in the TrOC removal.[43]

5.6.2 Redox Conditions and Chemical Structure

Microbial activity, diversity and the enzymatic range may all be boosted in treatment processes functioning under differing redox situations, all of which have a strong impact on microbial communities' TrOC degradation capacity. Some TrOCs (for example, roxithromycin, naproxen, erythromycin,

17-ethynylestradiol, and ibuprofen) could only be removed completely in aerobic circumstances. From a study conducted by Joss *et al.* (2004) it was seen that estrone and 17-ethynylestradiol were destroyed faster in aerobic environments, while tonalide and galaxolide were removed in both anoxic and aerobic conditions.[44] From a different investigation it was inferred that the anoxic situations are beneficial for the biodegradation of a range of TrOCs. The layout of the MBR compartments can potentially have an effect on TrOCs deterioration. The extent to which TrOCs degrade is also heavily influenced by their molecular structure and physical properties. The link between complex chemical structures and removal efficiency in treatment facilities has been the subject of numerous studies. Moreover, PHAs along with a comparable backbone and hydrophobic nature were capable of biodegrading specific compounds. Diclofenac, dichlorprop and clofibric acid, were all partially eliminated in the treatment systems due to the availability of chlorine atoms in their structures.[45] Other studies demonstrated the elimination of certain TrOCs employing enriched nitrifier culture and determined that compounds with chlorine atoms in their chemical structures had limited removal efficiency.[46,47] Although it is difficult to find a correlation between the efficiency of elimination and molecular structure, it was concluded that: (a) polycyclic substances and substances with electron withdrawing functional groups had low biodegradation; (b) TrOCs with electron donating functional groups had higher degradation efficiency; and (c) there were no substantial variations in the efficiencies of degradation of non-heterocyclic and heterocyclic chemicals.[48] From the investigations done by Sipma *et al.* (2010), it was concluded that the TrOCs with functional groups have increased biodegradability, but those containing halogen, iodine, and aromatic amine groups have reduced biodegradability.[49] Overall, these data show that the molecular properties and characteristics of selective enzymes involved in removal may have an important impact on the biodegradation efficiency.

5.6.3 Activated Sludge and Membrane Bioreactor Assisted Processes

There is enough evidence to support the chosen technology's impact on TrOC eradication (*e.g.* xenobiotic compounds). To study the rates of biodegradation of several TrOCs, Miege *et al.* (2008) employed a number of techniques, such as AS with and without N treatment, pre-treatment plus primary treatment, MBRs and others.[50] The AS technique implementing nutrient degradation seems to be among the highly efficient processes for the biodegradation of TrOCs. However, as compared to conventional technologies operating under the same operational conditions, it was not clear how these strategies facilitated the TrOC breakdown. Municipal and industrial wastewaters have been successfully treated with MBRs, which combine a suspended biomass biological reactor with a membrane microfiltration or ultrafiltration step. Despite their benefits (for example, smaller pore diameters that result in superior effluent), the majority of TrOCs aren't destroyed by size-exclusion methodologies. Another area that needs more research is the link between the cake layer and TrOC that forms on the surface of the membrane, because this layer can remove TrOCs *via* sorption/size exclusion.[51] An ultrafiltration MBR was found to destroy diclofenac and roxithromycin more successfully than a microfiltration MBR, with the differences related to cake layer retention. The implementation of MBRs instead of CAS in WWTPs is a straightforward approach of controlling biomass composition, allowing for the enrichment of slowly developing AOB and is projected to increase compound removal. In an investigation conducted by Weiss and Cozzarelli (2008) on the efficacy of AS and MBRs, employing MBRs significantly boosted the biodegradation efficiencies of 20 hydrophilic chemicals by a minimum of 25%.[52] The rise in the concentrations of the sludge and SRT rather than HRT may be attributed to the improved efficiency. According to a certain study, both methods (MBR and AS) produced comparable results in terms of TrOC eradication when used in similar contexts. Similarly, the removal of caffeine, bisphenol A, acetaminophen, and carbamazepine, ibuprofen, erythromycin, lincomycin and iopromide, by MBR and CAS processes demonstrated no crucial variations.[53,54] Other elements in both the systems could be essential. AOB abundance and community structure are influenced by process operation regimes, sewage parameters and reactor types. Rowan *et al.* (2003) investigated the community structure and diversity of the beta-proteobacterial AOB in two full-scale reactors that handled similar effluent.[55] A trickling filter contains more AOB variation than a biologically aerated filter, according to these researchers. A moving bed biofilm reactor, in addition to the methods listed above, has recently been discovered to exhibit biodegradation efficiency that is comparable to or better than MBRs or AS. The biggest downside of this technique, however, may be the decline in sludge settleability. Table 5.1 shows the effect of operating parameters on TrOC biodegradation and microbial activity.

5.7 Conclusion and Future Prospects

Several interpretations may be drawn from this chapter, which are summarized below: AOB have been identified in almost all sorts of environments, including extreme habitats such as terrestrial thermal/hot springs. Most TrOCs degrade primarily by biodegradation, which can occur *via* two crucial pathways *viz.* metabolism and co-metabolism. The aerobic cometabolic approach is the most common pathway for the degradation of TrOCs. Autotrophs (AOB) control the cometabolic biotransformation of TrOCs *via* AMO, while heterotrophs operate on TrOCs *via* both cometabolic and metabolic pathways. The successful biodegradation of TrOCs is highly dependent on understanding the microbial diversity (heterotrophic bacteria and AOB), the existence of a growth substrate *viz.* NH_4, and

Table 5.1 Effect of different parameters on TrOCs biodegradation and microbial activity. Reproduced from ref. 6 with permission from Elsevier, Copyright 2019.

Parameter	Influence on TrOC biodegradation and microbial activity
Temperature (*T*)	*T* ↑ (*e.g.*, up to 45 °C), biodegradation rate ↓ Improved elimination of TrOCs might be attained at warmer temperature *T* between 16 and 26 °C, no significant differences in removal of E2 and EE2 *T* between 10 and 20 °C, no significant differences in removal of E1 and E2 *T* range of 10–35 °C, more stable removal of hydrophobic compounds than those under 45 °C *T* ↑ (*e.g.*, 35 °C) could cause harmful effects on the bacterial enzymes and microbial growth activity *T* at 21 °C, elimination efficiencies ↑ (*e.g.*, atenolol, ibuprofen, and sulfamethoxazole) *T* in the range 18–23 °C, positive effect on sulfamethoxazole and erythromycin degradation removal
SRT	SRTs ↑ from 15 to 65 days, removal efficiency ↑ (*e.g.*, clofibric acid, diclofenac, ketoprofen, mefenamic acid, naproxen, tetracycline, oxytetracycline, sulfamethoxazole, sulfadiazine, and ampicillin) SRT ↑ from 8 to 80 days, removal efficiency ↑ (*e.g.*, gemfibrozil, ketoprofen, clofibric acid and 17α-ethynylestradiol) SRT ↑ >20 days, removal efficiency ↑ (*e.g.*, 17α-ethynylestradiol) SRT ↑ >30 days, removal efficiency ↑ (*e.g.*, pharmaceutical compounds) SRT >30 days, AOB were dominant in the AS system SRTs ↑ up to 80 days, no significant removal of TrOCs
pH	pH below 6 and 5, removal efficiencies ↑ (*e.g.*, acidic pharmaceuticals) At acidic pH (≤5), elimination efficiencies ↑ (*e.g.*, ibuprofen, gemfibrozil, indomethacin, naproxen, ketoprofen, clofibric acid and diclofenac) At pH 5, removal efficiencies ↑ (*e.g.*, sulfamethoxazole, ibuprofen, ketoprofen, and diclofenac) At lower acidic condition (*e.g.*, pH 4.5), impact microbial community structure involved in elimination of TrOCs
Redox conditions	Under redox conditions (*i.e.* anaerobic, anoxic, and aerobic), imp. the development of microbial populations and elimination efficiency of TrOCs. Aerobic condition – microbial diversity and activity of the microorganisms ↑ Under aerobic and anoxic conditions – TrOC elimination ↑ (*e.g.*, sulfamethoxazole) In both aerobic and anoxic conditions, TrOCs (*e.g.*, galaxolide and tonalide) could be removed In anoxic condition, removal efficiencies ↑ (*e.g.*, Diuron, tonalide, galaxolide and trimethoprim)

Table 5.1 (*Continued*)

Parameter	Influence on TrOC biodegradation and microbial activity
ASP and MBR related technologies	ASP are not effective in elimination of persistent/toxic TrOCs MBR can degrade most TrOCs (*e.g.*, persistent) MBR and AS technologies are similar for studied TrOCs ASP maybe effective in the biodegradation of easily biodegradable TrOCs (*e.g.*, ibuprofen, methyl paraben, galaxolide, and caffeine Superior biodegradation of naproxen, diclofenac, ketoprofen, gemfibrozil, propyphenazone, fluoxetine, ofloxacin, atenolol, propranolol, trimethoprim, lincomycin, tetracycline, *etc.*) was observed in MBR systems compared to that in CAS systems

the primary biotic and abiotic factors that influence the degradation of TrOCs throughout treatment procedures. Furthermore, to obtain comprehensive knowledge on the role of AOB in the biodegradation of TrOCs throughout the biological treatment process, it is essential to address the following issues.

- Further study is needed to identify the primary biodegradation mechanisms (*e.g.*, cometabolic or metabolic approach) and the formation of hazardous transformation products throughout TrOC biodegradation, with a special focus on the toxicity of specific TrOCs in comparison to their parent compounds.
- TrOCs are degraded by both autotrophic and heterotrophic bacteria. However, further research is required to understand the specific functions of these species and their enzymes (*e.g.*, dioxygenases and ammonia monooxygenase) involved in cometabolism and/or metabolism pathways.
- While there has been noticeable progress in the characterization and isolation of several species during nitrification, the activity and contribution of AOB under different environmental conditions, such as acidic soil habitats and hot springs, have received less attention and require further investigation.
- Numerous attempts have been made to improve the TrOC biodegradation *via* AOB. Nevertheless, most of the efforts are restricted to laboratory studies, and full-scale or pilot-scale trials are much needed to establish the performance efficiency and practicality of these systems.

References

1. H. A. Duong, N. H. Pham, H. T. Nguyen, T. T. Hoang, H. V. Pham, V. C. Pham, M. Berg, W. Giger and A. C. Alder, Occurrence, fate and

antibiotic resistance of fluoroquinolone antibacterials in hospital wastewaters in Hanoi, Vietnam, *Chemosphere*, 2008, **72**, 968–973.
2. B. Petrie, R. Barden and B. Kasprzyk-Hordern, A review on emerging contaminants in wastewaters and the environment: Current knowledge, understudied areas and recommendations for future monitoring, *Water Res.*, 2015, **72**, 3–27.
3. F. Ge, L. Y. Zhou, Y. Wang, Y. Ma, S. Zhai, Z. H. Liu, Y. J. Dai and S. Yuan, Hydrolysis of the neonicotinoid insecticide thiacloprid by the N2-fixing bacterium Ensifer meliloti CGMCC 7333, *Int. Biodeterior. Biodegrad.*, 2014, **93**, 10–17.
4. N. H. Tran, J. Hu and S. L. Ong, Simultaneous determination of PPCPs, EDCs, and artificial sweeteners in environmental water samples using a single-step SPE coupled with HPLC-MS/MS and isotope dilution, *Talanta*, 2013, **113**, 82–92.
5. X. Xu, X. Chen, P. Su, F. Fang and B. Hu, Biodegradation potential of polycyclic aromatic hydrocarbons by bacteria strains enriched from Yangtze River sediments, *Environ. Technol.*, 2016, **37**, 513–520.
6. M. Nsenga and F. Meng, Roles of ammonia-oxidizing bacteria in improving metabolism and cometabolism of trace organic chemicals in biological wastewater treatment processes: A review, *Sci. Total Environ.*, 2019, **659**, 419–441.
7. H.-D. Park, J. M. Regan and D. R. Noguera, Molecular analysis of ammonia-oxidizing bacterial populations in aerated-anoxic Orbal processes, *Water Sci. Technol.*, 2002, **46**, 273–280.
8. J. You, A. Das, E. M. Dolan and Z. Hu, Ammonia-oxidizing archaea involved in nitrogen removal, *Water Res.*, 2009, **43**, 1801–1809.
9. L. Van Niftrik, W. J. C. Geerts, E. G. Van Donselaar, B. M. Humbel, R. I. Webb, J. A. Fuerst, A. J. Verkleij, M. S. M. Jetten and M. Strous, Linking ultrastructure and function in four genera of anaerobic ammonium-oxidizing bacteria: Cell plan, glycogen storage, and localization of cytochrome c proteins, *J. Bacteriol.*, 2008, **190**, 708–717.
10. B. Kartal, J. Rattray, L. A. van Niftrik, J. van de Vossenberg, M. C. Schmid, R. I. Webb, S. Schouten, J. A. Fuerst, J. S. Damsté, M. S. M. Jetten and M. Strous, Candidatus 'Anammoxoglobus propionicus' a new propionate oxidizing species of anaerobic ammonium oxidizing bacteria, *Syst. Appl. Microbiol.*, 2007, **30**, 39–49.
11. B. Kartal, W. J. Maalcke, N. M. de Almeida, I. Cirpus, J. Gloerich, W. Geerts, H. J. M. Op den Camp, H. R. Harhangi, E. M. Janssen-Megens, K.-J. Francoijs, H. G. Stunnenberg, J. T. Keltjens, M. S. M. Jetten and M. Strous, Molecular mechanism of anaerobic ammonium oxidation, *Nature*, 2011, **479**, 127–130.
12. M. Oshiki, M. Ali, K. Shinyako-Hata, H. Satoh and S. Okabe, Hydroxylamine-dependent anaerobic ammonium oxidation (anammox) by 'Candidatus Brocadia sinica', *Environ. Microbiol.*, 2016, **18**, 3133–3143.

13. S. Gilch, O. Meyer and I. Schmidt, Electron paramagnetic studies of the copper and iron containing soluble ammonia monooxygenase from Nitrosomonas europaea, *Biometals*, 2010, **23**, 613–622.
14. N. Yamamoto, K. Otawa and Y. Nakai, Diversity and Abundance of Ammonia-Oxidizing Bacteria and Ammonia-Oxidizing Archaea During Cattle Manure Composting, *Microb. Ecol.*, 2010, **60**, 807–815.
15. D. E. Helbling, D. R. Johnson, M. Honti and K. Fenner, Process Parameters and Microbial Community Characteristics, *Environ. Sci. Technol.*, 2012, **46**, 10579–10588.
16. I. M. Head, W. D. Hiorns, T. M. Embley, A. J. McCarthy and J. R. Saunders, The phylogeny of autotrophic ammonia-oxidizing bacteria as determined by analysis of 16S ribosomal RNA gene sequences, *J. Gen. Microbiol.*, 1993, **139**, 1147–1153.
17. J. H. Rotthauwe, K. P. Witzel and W. Liesack, The ammonia monooxygenase structural gene amoa as a functional marker: Molecular fine-scale analysis of natural ammonia-oxidizing populations, *Appl. Environ. Microbiol.*, 1997, **63**, 4704–4712.
18. Å. Aakra, J. B. Utåker, A. Pommerening-Röser, H. P. Koops and I. F. Nes, Detailed phylogeny of ammonia-oxidizing bacteria determined by rDNA sequences and DNA homology values, *Int. J. Syst. Evol. Microbiol.*, 2001, **51**, 2021–2030.
19. E. Fernandez-Fontaina, F. Omil, J. M. Lema and M. Carballa, Influence of nitrifying conditions on the biodegradation and sorption of emerging micropollutants, *Water Res.*, 2012, **46**, 5434–5444.
20. H. Roh, N. Subramanya, F. Zhao, C. P. Yu, J. Sandt and K. H. Chu, Biodegradation potential of wastewater micropollutants by ammonia-oxidizing bacteria, *Chemosphere*, 2009, **77**, 1084–1089.
21. B. De Gusseme, B. Pycke, T. Hennebel, A. Marcoen, S. E. Vlaeminck, H. Noppe, N. Boon and W. Verstraete, Biological removal of 17α-ethinylestradiol by a nitrifier enrichment culture in a membrane bioreactor, *Water Res.*, 2009, **43**, 2493–2503.
22. J. Park, N. Yamashita, G. Wu and H. Tanaka, Removal of pharmaceuticals and personal care products by ammonia oxidizing bacteria acclimated in a membrane bioreactor: Contributions of cometabolism and endogenous respiration, *Sci. Total Environ.*, 2017, **605–606**, 18–25.
23. Y. Xu, Z. Yuan and B. J. Ni, Biotransformation of pharmaceuticals by ammonia oxidizing bacteria in wastewater treatment processes, *Sci. Total Environ.*, 2016, **566–567**, 796–805.
24. T. Alvarino, S. Suarez, J. M. Lema and F. Omil, Understanding the removal mechanisms of PPCPs and the influence of main technological parameters in anaerobic UASB and aerobic CAS reactors, *J. Hazard. Mater.*, 2014, **278**, 506–513.
25. J. Shi, S. Fujisawa, S. Nakai and M. Hosomi, Biodegradation of natural and synthetic estrogens by nitrifying activated sludge and ammonia-oxidizing bacterium Nitrosomonas europaea, *Water Res.*, 2004, **38**, 2323–2330.

26. M. Rattier, J. Reungoat, J. Keller and W. Gernjak, Removal of micropollutants during tertiary wastewater treatment by biofiltration: Role of nitrifiers and removal mechanisms, *Water Res.*, 2014, **54**, 89–99.
27. D. Janke and W. Fritsche, Nature and significance of microbial cometabolism of xenobiotics, *J. Basic Microbiol.*, 1985, **25**, 603–619.
28. N. H. Tran, T. Urase, H. H. Ngo, J. Hu and S. L. Ong, Insight into metabolic and cometabolic activities of autotrophic and heterotrophic microorganisms in the biodegradation of emerging trace organic contaminants, *Bioresour. Technol.*, 2013, **146**, 721–731.
29. J. B. Quintana, S. Weiss and T. Reemtsma, Pathways and metabolites of microbial degradation of selected acidic pharmaceutical and their occurrence in municipal wastewater treated by a membrane bioreactor, *Water Res.*, 2005, **39**, 2654–2664.
30. L. Iasur-Kruh, Y. Hadar and D. Minz, Isolation and bioaugmentation of an estradiol-degrading bacterium and its integration into a mature biofilm, *Appl. Environ. Microbiol.*, 2011, **77**, 3734–3740.
31. R. W. Murdoch and A. G. Hay, Formation of catechols *via* removal of acid side chains from ibuprofen and related aromatic acids, *Appl. Environ. Microbiol.*, 2005, **71**, 6121–6125.
32. L. Delgadillo-Mirquez, L. Lardon, J. P. Steyer and D. Patureau, A new dynamic model for bioavailability and cometabolism of micropollutants during anaerobic digestion, *Water Res.*, 2011, **45**, 4511–4521.
33. T. Yi and W. F. Harper, The link between nitrification and biotransformation of 17α-ethinylestradiol, *Environ. Sci. Technol.*, 2007, **41**, 4311–4316.
34. A. B. Hooper, T. Vannelli, D. J. Bergmann and D. M. Arciero, Enzymology of the oxidationof ammonia to nitrate by bacteria, *Antonie van Leeuwenhoek*, 1997, **71**, 59–67.
35. D. G. Wahman, A. E. Henry, L. E. Katz and G. E. Speitel, Cometabolism of trihalomethanes by mixed culture nitrifiers, *Water Res.*, 2006, **40**, 3349–3358.
36. C. P. Czajka and K. L. Londry, Anaerobic biotransformation of estrogens, *Sci. Total Environ.*, 2006, **367**, 932–941.
37. E. Kassotaki, G. Buttiglieri, L. Ferrando-Climent, I. Rodriguez-Roda and M. Pijuan, Enhanced sulfamethoxazole degradation through ammonia oxidizing bacteria co-metabolism and fate of transformation products, *Water Res.*, 2016, **94**, 111–119.
38. W. O. Khunjar, S. A. MacKintosh, J. Skotnicka-Pitak, S. Baik, D. S. Aga and N. G. Love, Elucidating the relative roles of ammonia oxidizing and heterotrophic bacteria during the biotransformation of 17α-ethinylestradiol and trimethoprim, *Environ. Sci. Technol.*, 2011, **45**, 3605–3612.
39. Y. Xu, J. Radjenovic, Z. Yuan and B. J. Ni, Biodegradation of atenolol by an enriched nitrifying sludge: Products and pathways, *Chem. Eng. J.*, 2017, **312**, 351–359.
40. N. H. Tran, T. Urase and O. Kusakabe, The characteristics of enriched nitrifier culture in the degradation of selected pharmaceutically active compounds, *J. Hazard. Mater.*, 2009, **171**, 1051–1057.

41. D. J. Arp, C. M. Yeager and M. R. Hyman, Molecular and cellular fundamentals of aerobic cometabolism of trichloroethylene, *Biodegradation*, 2001, **12**, 81–103.
42. L. Yang, Y. F. Chang and M. S. Chou, Feasibility of bioremediation of trichloroethylene contaminated sites by nitrifying bacteria through cometabolism with ammonia, *J. Hazard. Mater.*, 1999, **69**, 111–126.
43. J. S. Vader, C. G. Van Ginkel, F. M. G. M. Sperling, J. De Jong, W. De Boer, J. S. De Graaf, M. Van Der Most and P. G. W. Stokman, Degradation of ethinyl estradiol by nitrifying activated sludge, *Chemosphere*, 2000, **41**, 1239–1243.
44. A. Joss, H. Andersen, T. Ternes, P. R. Richle and H. Siegrist, Removal of estrogens in municipal wastewater treatment under aerobic and anaerobic conditions: Consequences for plant optimization, *Environ. Sci. Technol.*, 2004, **38**, 3047–3055.
45. K. Kimura, H. Hara and Y. Watanabe, Removal of pharmaceutical compounds by submerged membrane bioreactors (MBRs), *Desalination*, 2005, **178**, 135–140.
46. N. Tadkaew, F. I. Hai, J. A. McDonald, S. J. Khan and L. D. Nghiem, Removal of trace organics by MBR treatment: The role of molecular properties, *Water Res.*, 2011, **45**, 2439–2451.
47. Y. Luo, W. Guo, H. H. Ngo, L. D. Nghiem, F. I. Hai, J. Zhang, S. Liang and X. C. Wang, A review on the occurrence of micropollutants in the aquatic environment and their fate and removal during wastewater treatment, *Sci. Total Environ.*, 2014, **473–474**, 619–641.
48. A. Joss, S. Zabczynski, A. Göbel, B. Hoffmann, D. Löffler, C. S. McArdell, T. A. Ternes, A. Thomsen and H. Siegrist, Biological degradation of pharmaceuticals in municipal wastewater treatment: Proposing a classification scheme, *Water Res.*, 2006, **40**, 1686–1696.
49. J. Sipma, B. Osuna, N. Collado, H. Monclús, G. Ferrero, J. Comas and I. Rodriguez-Roda, Comparison of removal of pharmaceuticals in MBR and activated sludge systems, *Desalination*, 2010, **250**, 653–659.
50. C. Miège, J. M. Choubert, L. Ribeiro, M. Eusèbe and M. Coquery, Removal efficiency of pharmaceuticals and personal care products with varying wastewater treatment processes and operating conditions – conception of a database and first results, *Water Sci. Technol.*, 2008, **57**, 49–56.
51. Y. Wu, X. Ke, M. Hernández, B. Wang, M. G. Dumont, Z. Jia and R. Conrad, Autotrophic growth of bacterial and archaeal ammonia oxidizers in freshwater sediment microcosms incubated at different temperatures, *Appl. Environ. Microbiol.*, 2013, **79**, 3076–3084.
52. J. V. Weiss and I. M. Cozzarelli, Biodegradation in contaminated aquifers: Incorporating microbial/molecular methods, *Ground Water*, 2008, **46**, 305–322.
53. M. Clara, N. Kreuzinger, B. Strenn, O. Gans and H. Kroiss, The solids retention time – A suitable design parameter to evaluate the capacity of

wastewater treatment plants to remove micropollutants, *Water Res.*, 2005, **39**, 97–106.
54. J. Radjenović, M. Petrović and D. Barceló, Fate and distribution of pharmaceuticals in wastewater and sewage sludge of the conventional activated sludge (CAS) and advanced membrane bioreactor (MBR) treatment, *Water Res.*, 2009, **43**, 831–841.
55. A. K. Rowan, G. Moser, N. Gray, J. R. Snape, D. Fearnside, T. P. Curtis, M. R. Barer and I. M. Head, A comparitive study of ammonia-oxidizing bacteria in lab-scale industrial wastewater treatment reactors, *Water Sci. Technol.*, 2003, **48**, 17–24.

CHAPTER 6

The Heterotrophic Nitrification and Aerobic Denitrification (HN–AD) Process

G. ANJALI AND P. C. SABUMON*

School of Civil Engineering, Vellore Institute of Technology, Chennai-600127, Tamil Nadu, India, Email: anjaligopakumar@vit.ac.in
*Email: pcsabumon@vit.ac.in

6.1 Introduction

Biological nitrogen removal (BNR) processes are effective processes[1] to reduce the reactive nitrogen (NH_4–N, NO_2–N and NO_3–N) load in the environment *via* an eco-friendly route.[2] The traditional BNR process consists of a two-step process initiated by an aerobic-autotrophic nitrification and anoxic-heterotrophic denitrification process. These two steps are facilitated by physiologically distinct microbial species. The first step of nitrification sequentially converts the NH_4–N to NO_2–N and then to NO_3–N in the presence of sufficient concentrations of oxygen and specialized microbial species, Ammonium Oxidizing Bacteria (AOB) and Nitrite Oxidizing Bacteria (NOB), respectively. This step is considered the rate limiting step of the BNR process. The NO_3–N formed is then degraded to nitrogen gas by the anoxic–heterotrophic denitrifiers in the presence of organic matter as an electron donor.

The major limitations of the traditional BNR process include the oxygen requirement for the nitrification process and the organic carbon requirement for the denitrification process. This renders the traditional BNR process more energy intensive and costly. Many researchers[3] have demonstrated

Ammonia Oxidizing Bacteria: Applications in Industrial Wastewater Treatment
Edited by Maulin P. Shah

Published by the Royal Society of Chemistry, www.rsc.org

the development of partial nitrification (restriction of oxidation until formation of NO_2–N) at low concentrations of DO and the occurrence of denitrification at low organic carbon contents. But still many challenges prevail in the process, such as: (a) the need for high solid retention time, (b) regulation of temperature, alkalinity, pH and inhibitory compounds, (c) the need of a large footprint, and (d) production of excess sludge.[4] These challenges prompt the need for (i) advances in the explorations of novel and innovative microbial processes for sustainable nitrogen removal and (ii) developing various cost-effective biotechnological treatment methods to meet the stringent discharge standards as low as 10 mg N per L set by many countries.[5]

6.1.1 New Nitrogen Removal Processes

The type of novel microbial processes and the corresponding type of reactor system for nutrient removal mostly will vary with the influent COD to Nitrogen (C/N) ratio. Many autotrophic processes such as anaerobic ammonium oxidation (ANAMMOX), single high-rate ammonia removal over nitrite (SHARON), complete autotrophic nitrogen removal over nitrite (CANON), oxygen limited autotrophic nitritation/denitrification (OLAND), denitrifying ammonium oxidation (DEAMOX), *etc.*, are reported to be autotrophic processes tolerant of low organic carbon concentrations. They efficiently treat wastewaters characterized with low C/N ratio.[6] These autotrophic processes solve the problems of incomplete denitrification in traditional BNR in the presence of insufficient organic carbon and are considered as sustainable nitrogen elimination technologies.

Other recent nitrogen removal processes include microbial consortiums classified as ammonium oxidizing archaea (AOA). They are reported to have the ammonium oxidizing enzyme called ammonium mono oxygenase (AMO), like in AOBs. This enables them to play an active role in catalysing ammonia oxidation.[7] Recently, complete ammonia oxidation (COMAMMOX) was reported to be catalyzed by a single microorganism, *Nitrospira* sp. They adapt better to conditions of low DO.[8] Many studies have been reported on the established Anammox process in the presence of various electron acceptors like NO_2–N/NO_3–N, ferric oxides (FEAMMOX),[9] sulphates, organic compounds (anaerobic ammonium oxidation using *Candidatus fulgidae*), *etc.* The anammox process proved to have a wide range of applications in treating complex wastewaters characterized with low C/N ratio. Several hybrid processes such as simultaneous nitrification anammox and denitrification (SNAD), simultaneous partial nitrification and Anammox (SNAP) or PN–A, simultaneous anammox and denitrification (SAD), *etc.* consisting of the coexistence of different nitrogen removal species have been developed for treating wastewater characterized with $C/N \leq 2$. The basic pathways mediated by functional enzymes for nitrogen removal processes used in wastewater treatment are shown in eqn (6.1)–(6.7). The major functional enzymes for nitrogen transformation are ammonia mono oxygenase (AMO), hydroxylamine oxidoreductases (HAO or *hao*), nitrite oxidoreductases (NOR or *Nor*), nitrite reductase (NIR or *NiR*), nitrate reductases (NXR), nitrous oxide reductases (NOS), cytochrome *c* nitrate reductases (ccNIR), NCyA (uncharacterized enzyme converting NO to NO_2^-) and hydrazine hydroxylase (HH).

(i) Autotrophic nitrification mediated by ammonia oxidizing bacteria (AOB):

$$NH_3 \xrightarrow{\text{AMO}} NH_2OH \xrightarrow{\text{HAO}} NO \xrightarrow{\text{NCyA}} NO_2^- \xrightarrow{\text{NXR (NOB)}} NO_3^- \tag{6.1}$$

(ii) Reciprocal feeding pathway mediated by urease positive *Nitrospira moscoviensis* (NOB):

$$\text{Urease} \xrightarrow{\text{NOB}} NH_3 \xrightarrow[\ ]{\substack{\text{AOB} \\ \text{(Urease } -\text{ve, Nitrosomonas europea)}}} NO_2^- \xrightarrow{\text{NOB}} NO_3^- \tag{6.2}$$

(iii) Comammox process mediated by *Nitrospira* sp.:

$$NH_4^+ \xrightarrow{\substack{\textit{Nitrospira}\ \text{sp.} \\ \text{AMO}}} NH_2OH \xrightarrow{\text{HAO}} NO_2^- \xrightarrow{\text{NXR}} NO_3^- \tag{6.3}$$

(iv) Heterotrophic denitrification:

$$NO_3^- \xrightarrow{\frac{\text{NAR}}{\text{NAP}}} NO_2^- \xrightarrow{\text{NIR}} NO \xrightarrow{\text{NOR}} N_2O \xrightarrow{\text{NOS}} N_2 \tag{6.4}$$

(v) Simultaneous nitrification and denitrification (SND):

$$NH_4^+ \xrightarrow{\text{AMO}} NH_2OH \xrightarrow{\substack{\text{HAO} \\ (O_2)}} NO \xrightarrow{\text{NCyA}} NO_2^- \xrightarrow{\text{NIR}} NO \xrightarrow{\text{NOR}} N_2O \xrightarrow{\text{NOS}} N_2 \tag{6.5}$$

(vi) Dissimilatory nitrate reduction to ammonia:

$$NO_3^- \xrightarrow{\text{NXR}} NO_2^- \xrightarrow{\text{ccNIR}} NH_4^+ \tag{6.6}$$

(vii) Anammox process mediated by anaerobic anammox bacteria:

$$\begin{array}{l} NH_4\text{-}N \searrow \\ NO_2^- \xrightarrow{\text{NIR}} NO \xrightarrow{\text{HH}} N_2H_2 \xrightarrow{\text{HAO}} N_2 \end{array} \tag{6.7}$$

When the concentration of organic carbon becomes very prominent in effluents from the pharmaceutical industry, tanneries, livestock or slaughterhouse waste processing, potato processing industries, alcohol and starch production, formaldehyde production and palm oil mill effluent, *etc.*, the autotrophic processes or the hybrid processes fail to achieve efficiency due to the process and microbial limitations/inhibitions.[10] In the treatment of such wastewaters, we require the nitrification to happen heterotrophically or simultaneous nitrification and denitrification (SND) in a single reactor.

6.1.2 Heterotrophic Nitrification and Aerobic Denitrification (HN–AD)

The concept of oxidizing ammonia heterotrophically to nitrogen in the presence of an abundance of oxygen is known as Heterotrophic Nitrification and Aerobic Denitrification (HN–AD). Multiple pathways are proposed based on the diversity and functional genes or enzyme activities available in the species. The feasible pathways followed by most of the HN–AD microbes were explained by Chen and Ni[11] and Huang *et al.*[12] In one pathway, the process of heterotrophic nitrification happens *via* oxidation of ammonium to nitrite and thereafter the nitrite is converted to nitrogen gas resulting in aerobic denitrification ($NH_4^+ \rightarrow NH_2OH \rightarrow NO_2^- \rightarrow N_2O \rightarrow N_2$).[11] The other pathway is hydroxylamine transformation to N_2O and/or N_2 rather than nitrite ($NH_4^+ \rightarrow NH_2OH \rightarrow N_2O \rightarrow N_2$).[12] This is a counter strategy to the conventional BNR process. This becomes a favorable strategy because it can result in simultaneous nitrification and denitrification (SND) in a single reactor.

The major attribute of HN–AD species is the coexistence of enzymes AMO and periplasmic nitrate reductase (NAP) in a single bacterium. These enzymes help in transferring NH_4–N to NO_2–N/NO_3–N and further converting it to N_2 independently. This avoids the competition between heterotrophic denitrifying bacteria and autotrophic nitrifying bacteria for substrates and intermediate products. Also, HN–AD bacteria have a rapid growth rate, which will be beneficial for the rapid startup and stability of the reactor.[13]

The process was first observed by Robertson *et al.*[14] in the early 1980s. They reported that *Thiosphaera pantotropha* sp. was capable of heterotrophic nitrification and aerobic denitrification (HN–AD) processes. From then on, many HN–AD species were observed, isolated and developed from various natural environments such as soil environments, surface water (lakes and rivers), activated sludge, piggery/swine wastewater, landfill leachates, coking plant wastewaters, pharmaceutical wastewaters, saline wastewaters, construction wetlands, Abattoir wastewater, *etc.* The wide range of sources from which HN–AD species can be enriched shows the ubiquitous nature of the species.[15] These species are reported innately with many advantages[15] like: (1) rapid growth rate and biomass yield compared to autotrophic nitrifiers; (2) improved alkalinity as complete SND happens in a single reactor; (3) high efficiency and energy savings over conventional denitrification methods,[16] (4) simple process design and operational conditions; (5) high tolerance to environmental stress such as salinity, heavy metal ions and antibiotics.[17,18]

All of these advantages and the versatility exhibited by HN–AD species make it worth employing them for real wastewaters with complex characteristics.

For advancing the HN–AD process for wastewater treatment, knowledge of diversity of the biomass, the major bio-chemical pathways and environmental factors regulating the reactions are essential. This chapter is organized to review the existing literature on the diversity of HN–AD species, their habitats, their identification, and isolation techniques. The chapter also talks about the various proposed pathways coordinated by HN–AD microbes for nutrient removal. Emphasis is also given to the environmental factors affecting the performance of HN–AD microbes. The chapter ends with various applications and strategies used to improve the application of the HN–AD process and future research required in the field of heterotrophic nitrification and aerobic denitrification.

6.2 Microbes Responsible for the HN–AD Process

The application of the HN–AD process is considered as a more promising strategy than the traditional BNR process when treating wastewaters, especially when the COD to NH_4–N ratio is high and at high concentration of the substrates (COD and NH_4–N; C/N ratio ≥ 2). During this scenario, the conventional nitrification process efficiency gets affected due to the slow growth of the autotrophic bacteria and its inhibition. Many recent studies have also reported the capability of HN–AD microbes to nitrify and denitrify simultaneously even when exposed to severe conditions (*i.e.*, acidic pH, high salinity, psychrophilic condition, presence of toxic compounds, and so on).[19] Many species can also exhibit either heterotrophic nitrification or aerobic denitrification alone. However, in this chapter, more emphasis is given to those species capable of undergoing heterotrophic nitrification and aerobic denitrification processes together.

6.2.1 Diversity of HN–AD Microbes

The HN–AD microbes are distributed extensively and are found to withstand different environmental conditions. The autotrophic nitrifying species are normally classified as Gram negative bacteria, while HN–AD species consists of Gram negative and Gram-positive bacteria and fungi.[20] To date we have more than 50 species of HN–AD microbes classified under the phylum Proteobacteria (alpha, beta, and gamma), Firmicutes, Fungi, Bacteroidota, and Actinobacteria. Major classifications of HN–AD species belong to the phylum Gamma Proteobacteria.[15,20] A comprehensive review of their diverse habitats and unique characteristics is represented in Table 6.1. As they are isolated from different habitats, they are considered physiologically diverse. Due to these different physiological characteristics, HN–AD microbes may be more versatile when compared to AOBs. The versatility and potential functions need to be researched more as earlier studies were mainly emphasized on cell growth, metabolic activities of microbes and nitrogen conversion rates.[21]

Table 6.1 Diversity of HN–AD microorganisms.

Phylum	Genus	Species	Sources	Unique feature	Ref.
Gram negative, gamma proteobacteria	*Pseudomonas*	*stutzeri YZN-001*	Swine manure effluent	Psychrotolerant HN–AD sp. removed NH_4– at low temperature.	15
	Pseudomonas	*stutzeri YG-24*	Wastewater treatment	Phosphate accumulating bacteria along HN–AD	23
	Pseudomonas	*putida NP5*	Activated sludge	Nitrogen assimilation and nitrogen dissimilation possible	24
	Pseudomonas	*tolaasii Y-11*	Paddy soil	Maintain stable growth and nitrogen removal effect when the nitrite reached 200 mg L^{-1}	25
	Pseudomonas	*mendocina X49*	Beijing sewage treatment plant	Rapidly degrade high concentrations of inorganic nitrogen (5 to 100 mg L^{-1})	26
	Pseudomonas	*bauzanensis DN13-1*	Deep sea sediment	Strain could effectively remove NO_2–N and NH_4–N. Functional genes identified include genes nirS, norB, nosZ, nasA and putative amo	21
	Pseudomonas	*balearica UFV3*	Wastewater treatment plant	Ability of complete nitrogen removal at C/N: 8, in presence of citrate as external carbon source, pH 7 to 9	27
	Pseudomonas	*ADN-42*	Marine sediments	Capable of efficient HN–AD at low DO $\approx$ 3 mg L^{-1}; high salinity tolerant strain	28
	Acinetobacter	*HA-2*	Pre enriched mixed culture	Psychrotolerant HN–AD sp. (Excellent tolerance to low temperature with 20 °C as its optimum and 4 °C as viable)	29
	Acinetobacter	*Y16*	River	Cold adapted strain able to achieve 66% of NH_4–N removal through HN–AD mechanism at C/N ratio 2	12
	Acinetobacter	*JR1*	Pharmaceutical wastewater	Acidophilic bacteria exhibited wide range of HN–AD activity at pH 4.5 to 10, optimum at pH 4	30

Acinetobacter	*baumannii AL-6*	Activated sludge	Efficient HN–AD with chromium(VI) tolerance	15
Acinetobacter	*T1*	Piggery wastewater	Used as the inoculum for the first field scale reactor reported	31
Acinetobacter	*TAC 1*	Swine wastewater	Psychrotolerant HN–AD species capable of the process at 5 °C	15
Acinetobacter	*junii YB*	Activated sludge	Exhibited capability of aggregation and hydrophobicity. pH range 4 to 11	32
Klebsiella	*pneumoniae Y6*	Seawater	Effective treatment of sea water based wastewater. Resistant to antibiotics	33
Klebsiella	*pneumoniae EGD-HP19-C*	Nitroaromatic industrial wastewater	The species possess incomplete nitrification and denitrification pathways. But still removes the inorganic nitrogen content from the system *via* ammonification reaction	34
Aeromonas	*HN-02*	Activated sludge	Halophilic strain, can withstand salinity of 20 $g L^{-1}$	15
Paracoccus	*pantotropha ATCC 35512*	Soil and sea sludge or wastewater treatment plant	Primitive species conducting HN–AN with prominence to nitrate reductase	20
Agrobacterium	*LAD9*	Landfill leachates	Succinate and acetate was the favorable carbon. The species could simultaneous perform dissimilatory and assimilatory reactions	15
Cupriavidus	*S1*	Coking plant wastewater	Metal resistant Bacteria, the bacteria is resistant to Zn^{2+}, Ni^{2+}, Cu^{2+}, Cr^{6+}	17
Zobellelle	*taiwanensis DN-7*	Landfill leachates	Treat high strength NH_4–N up to 2000 $mg L^{-1}$; removal rate of 17.3 $mg L^{-1} h^{-1}$ at C/N = 8	15

Table 6.1 (*Continued*)

Phylum	Genus	Species	Sources	Unique feature	Ref.
	Alcaligenes	*TB*	Biofilter biofilm	Reduction of nitrate in the presence of nanoscale zero-valent iron (nZVI)	35
	Alcaligenes	*faecalis WT14*	Constructed wetland	High NH_4–N tolerance (up to 2000 mg L^{-1} without inhibition), fast growth rate, strong NO_2–N/NO_3–N removal capacity	36
	Diaphorobacter	*PD7*	Coking plant wastewater	Capable of phenol degradation	37
	Diaphorobacter	*polyhydroxybutyrativorans SL-205*	Denitrified wastewater	Use poly(3-hydroxybutyrate-*co*-3-hydroxyvalerate as sole carbon source	15
	Ochrobactrum	*HND19*	Activated sludge	Effective removal of nitrogen species with removal rate 4.28 and 4.01 mg N per L per h	38
	Serratia	*marcescens W5*	Lake water	Alkaline resistant HN–AD sp. High NH_4–N removal at alkaline pH (9)	39
	Vibrio	*diabolicus SF16*	Marine sediments	Halophilic strain with a salinity adaptability range of 10 to 50 g L^{-1}	24
	Providencia	*rettgeri YL*	Wastewater	Inorganic ions such as Mg^{2+}, Mn^{2+}, and Zn^{2+} largely influenced the growth and nitrogen removal efficiency	40
	Rhodococcus	*CPZ24*	Swine wastewater	The ammonium nitrogen removal rate was 3.4 mg L^{-1} h^{-1}	41
	Bacillus	*methylotrophicus strain L7*	Wastewater	Halophilic strain, can withstand salinity of 30 g L^{-1}	24
	Bacillus	*litoralis N31*	Mariculture wastewater	Displayed efficient nitrification ability at pH 7.5 to 8.5	16
	Bacillus	*MS30*	Deep sea hydrothermal vent	Halophilic and thermophilic HN–AD, optimum at 65 °C and 25 g L^{-1} as salinity	28
	Enterobacter	*cloacae HW-15*	Phosphorus-rich wastewater	Capable of removing nutrients and arsenic(v) at an arsenic(v) concentration lower than 1.5 mg L^{-1}	22

	Stenotrophomonas	*maltophilia DQ 01*	Ubiquitous bacteria in aqueous, soil and plant environment	Adheres strongly and forms biofilm on plastic surfaces	13
	Stenotrophomonas	*MSNA-1*	High salinity and acidic/alkaline environment	Acid tolerant HN–AD microbe. Optimum pH: 7.5	42
Gram negative beta Proteobacteria	*Comamonas*	*denitrificans* sp.	Activated sludge	Aerobic denitrifier	43
	Microvirgula	*aerodenitrificans*	Activated sludge	Aerobic denitrifier	20
	Achromobacter	*GAD3*	Activated sludge in a landfill leachate treatment system	Aerobic denitrifier and phosphorous removal	43
Alpha proteo bacteria	*Pannonibacter*	*W30*	Highly polluted surface water	Assimilation rather than denitrification was the main mechanism for the environmental nitrogen	15
Firmicutes (Bacillota)	*Bacillus*	*licheniformis*	Soil	Mesophilic bacteria having optimum at 50 °C	44
	Bacillus	*simplex H-b*	Frozen soil	Psychrotolerant HN–AD species capable of the process at 10 °C	15
	Exiguobacterium	*mexicanum*	Saline wastewater	Halophilic strain	15
	Anoxybacillus	*contaminans HA*	Biofilter biofilm	Thermophilic HN–AD species	15
Fungi	*Aspergillus*	*luchuensis*	Municipal wastewater treatment	Multi and micro pollutant remover, phosphorous, nitrogen and organic carbon	45
	Penicillium	*tropicum ISO293*	Surface water	Enhanced nitrate removal over a variety of organic carbon components	20
	Penicillium	sp. *M25-22*	Chicken faeces compost	Capable of oxidizing NH_4–N in presence of glucose, sucrose, starch, or cellulose	20
	Trametes	*versicolor*	Laboratory isolate	Effective multi and micropollutant remover	45
	Barnettozyma	*californica K1*	Surface sediment	Able to utilize different carbon sources in the order of optimum carbon source sucrose > glucose > acetate > citrate	46

Table 6.1 (*Continued*)

Phylum	Genus	Species	Sources	Unique feature	Ref.
Bacteroidota	*Chryseobacterium*	*R31*	Abattoir wastewater	Functions independent to type of nitrogen source	47
Actinobacteria	*Arthrobacter*	*arilaitensis Y-10*	Soil and sewage	Can work at temperature 15 °C	15
	Arthrobacter	*globiformis*	Ubiquitous soil inhabitant	Hydroxylamine mediated nitrogen removal	48
	Gordonia	*amicalis UFV4*	Wastewater treatment plant	Ability of complete nitrogen removal at C/N: 8, in presence of citrate as external carbon source	27
Halomonadaceae	*Halomonas*	*BO1*	High salinity nitrogenous wastewater	Halophilic bacteria capable of HN–AD at pH ranging between 8 and 9	15

Huang *et al.*[16] reported that multiple genera of HN–AD microbes function differently in various conditions. For example, the genera *Klebsiella* classified as a HN–AD microbe is reported to cause diseases when infected, while genera *Bacillus* prevents diseases and improves water quality. Also, many known environmental microbes have been reported to exhibit the HN–AD process. For instance, *Enterobacter Cloacae*, known as a phosphorus accumulating bacteria (PAB), could achieve 99% NH_4–N removal, 59% NO_2–N removal and NO_3–N removal up to 88%, when employed to treat real sewage.[22] Hence, the potency of microbes performing the HN–AD process is highly diverse. This diversity has made more coexistence possible. Currently, hybrid processes which can have combinations of different mediating microbes are considered an appropriate solution for removing multiple pollutants in wastewater. Recently, a lab scale moving bed biofilm reactor (MBBR) was set up where a mixed culture was inoculated for treating high ammonia wastewater. The dominant species were found to be *Acinetobacter*, *Pseudomonas* and *Paracoccus*, along with other bacteria including *Halomonas* and *Thauera*. The coexistence improved the tolerance of the microbes to NH_4–N to the extent of 1000 $mg\,L^{-1}$. Even the startup time was less (15 days).[21]

6.2.2 Common Bio-augmentation and Identification Techniques

Employing the HN–AD process for treating nitrogen rich wastewaters, especially when it is characterized with high C/N ratio, has received tremendous attention. For this, bio-augmentation of HN–AD species is very necessary. The major steps towards bio-augmentation of microbes are (i) isolation of the required species, and (ii) culturing and enrichment of the required species. The HN–AD bio-augmentation has been successfully done using cultivation-dependent approaches where the above-mentioned steps are very important.[15]

The commonly used media for cell culturing or isolation of pure cultures are modified bromothymol blue (BTB) medium. The initial microbial consortium (environmental sample) is homogenized in saline solution followed by serial dilutions. The diluted samples are then streaked on modified BTB medium to grow pure culture colonies. The morphologically different colonies are further isolated and purified *via* repetitive streaking on the modified BTB medium.[29] The pure culture isolated would further go for rapid screening and enrichment. The enrichment mainly happens in various nutrient media feasible for the HN–AD process. The basic nutrient media used for the enrichment of HN–AD microbes include heterotrophic nitrification medium (HNM), nitrite denitrification medium (NDM) and nitrate denitrification medium (DM).[17] These media are used for increasing the population of the desired HN–AD microbes as well as used to test their capacity for removing NH_4–N, NO_2–N and NO_3–N from the wastewater/synthetic wastewater. Finally, the pure cultures depicting excellent performances are screened for subsequent bioassays and applications.[17] The in-depth studies to understand the general molecular information of HN–AD microbes using modern omics approaches (genomics, transcriptomics, and proteomics) are yet to be researched.

The metabolic activities of specific HN–AD microbes are generally recognized by amplifying the functional genes or enzymatic activities. The major marker enzymes used for tagging HN–AD microbes are *hao*, *napA*, *nirS*, and *nirK*. They are also used in identifying microbes responsible for the nitrite reduction processes. The diversity in HN–AD microbes is also understood based on the presence and the absence of these functional enzymes. For example, Xie *et al.*,[49] reported that the functional genes present in *Acinetobactor* sp. *ND7* were *hao*, *napA* and *nirS*, while *nirK* was not detected. The nitrite denitrification in the species was found to happen *via* a homodimer cytochrome *c*d1 nitrite reductase (cd1-NIR), giving rise to an alternative nitrogen removal pathway by the HN–AD microbe. Likewise specific assay studies are conducted to detect the major marker enzymes involved in the nitrogen transformation processes. The information on the functional enzymes is restricted to *hao*, *nxr*, *nar*, *nir* and *nor* genes.[32] Hence, more in-depth studies are needed to understand the specificity of each HN–AD species. Recently, the quorum sensing (QS) technique was used for the identification of HN–AD species. It helps to regulate gene expression in response to fluctuations in cell population density using signal molecules.[50]

6.3 Nutrient Removal Mechanisms Mediated by HN–AD Microbes

The metabolic pathways of HN–AD microbes are considered complex due to the variations in the presence of functional genes for different species. Many pathways are proposed and observed based on the enzymatic activities. It is found that[19] HN–AD species have the capacity for assimilation and dissimilation of nitrogen compounds. The major transformations between organic nitrogen, inorganic nitrogen and gaseous nitrogen are mediated by HN–AD microbes through oxidation, reduction, and assimilation reactions.

6.3.1 Nitrogen Dissimilation Mechanisms by HN–AD Microbes

Based on various hypotheses and observations, the mechanism of complete nitrogen removal/dissimilation by HN–AD microbes is classified into two routes as shown in eqn (6.8) and (6.9).

$$NH_4^+ \rightarrow NH_2OH \rightarrow NO_2^- \rightarrow N_2O \rightarrow N_2 \tag{6.8}$$

proposed by Chen and Ni,[11]

$$NH_4^+ \rightarrow NH_2OH \rightarrow N_2O \rightarrow N_2 \tag{6.9}$$

proposed by Huang *et al.*[12]

The above pathways result in complete assimilation reaction from inorganic nitrogen to gaseous nitrogen. For both of these pathways, organic carbon is used as the carbon source.

6.3.1.1 Nitrogen Removal Through the Nitrite–Nitrate Pathway

This model was the basic and widely accepted model proposed by Wehrfritz *et al.*[48] explaining the complete nitrogen removal by HN–AD microbes. The major enzymes mediating the heterotrophic nitrification reactions were ammonia monooxygenase (AMO), hydroxylamine oxidoreductases (HAO), nitrite oxidoreductases (NOR), pyruvic oxime dioxygenase (POD), and nitrate reductases (NXR). The pathway followed is as shown in eqn (6.10) and (6.11). The initial step consists of heterotrophic nitrification. The oxidation of hydroxylamine was observed to be mediated by either HAO or POD.

Heterotrophic Nitrification in the Presence of HAO

$$NH_4^+ \xrightarrow{AMO} NH_2OH \xrightarrow{HAO} NO_2^- \xrightarrow{NOR} NO_3^- \tag{6.10}$$

as proposed by Wehrfritz *et al.*[48] and Duan *et al.*[51]

Heterotrophic Nitrification in the Presence of POD

$$NH_4^+ \xrightarrow{AMO} NH_2OH \xrightarrow{POD} NO_2^- \xrightarrow{NOR} NO_3^- \tag{6.11}$$

as proposed by Tsujino *et al.*[52]

The AMO enzyme is characterized as an unstable membrane-bound protein, and HAO is known to be a multiheme *c* (high molecular weight enzyme). These enzymes are seen in AOBs also, which mediate the traditional BNR process. The HN–AD process proceeds in either of the two pathways based on the presence of the enzyme. Wehrfritz *et al.*[48] reported that the HN–AD species *Arthrobacter globiformis* and *Paracoccus pantotropha* catalyze the formation of hydroxylamine using AMO. But the oxidation of hydroxylamine was observed to be mediated by the enzyme POD, which is a non heme Fe protein. *Alcaligenes faecalis* also exhibits a similar metabolism where the oxidation of hydroxylamine was mediated by POD.[52] The enzyme converts hydroxylamine to pyruvic oxime by a non-enzymatic reaction with pyruvate. The pyruvic oxime is converted to pyruvate and nitrite during deoxygenation reaction by POD.

The pathway of nitrate–nitrite reduction to complete the HN–AD process is like heterotrophic denitrification. The major enzymes required for the reduction pathway are nitrate reductases (NXR) and nitrite reductase (NIR) as shown in eqn (6.12). Nitrate reductases consist of three distinct types: (i) periplasmic nitrate reductase (Nap), (ii) respiratory nitrate reductase (Nar) and (iii) assimilatory nitrate reductase (Nas). Under aerobic conditions, Nap and Nas convert nitrate to nitrite effectively indicating suitability for aerobic denitrification.[21,36] The final step is mediated by a copper-containing enzyme nitrite reductase (NirK) or cytochrome *c*d1 nitrite reductase (NirS). When the NO_2–N formed is directly converted to N_2, the reaction is mediated by NirK. When the NO_2–N is converted to nitric oxide and then to N_2, the reaction is mediated by NirS.[23]

Aerobic Denitrification Pathway

$$\rightarrow NO_3^- \xrightarrow{NXR} NO_2^- \xrightarrow{NIR} N_2 \tag{6.12}$$

as proposed by Wehrfritz *et al.*[48]

Even though this pathway is widely accepted, still more research needs to be done to understand the functional enzymes responsible. In a recent study,[21] a different characteristic was observed in a HN–AD species *Pseudomonas bauzanensis DN13-1*. This species was found to undergo the hydroxylamine conversion to nitrite pathway, but when verified for functional enzymes, both HAO and POD were absent. Hence, it was clear that there are chances of other pathways also for heterotrophic nitrification depicting insufficiency in current research on key enzymes.

6.3.1.2 Nitrogen Removal Through the Hydroxylamine Reduction Pathway

The enormous diversity in HN–AD microbes shows the feasibility of multiple pathways of reaction. The basic procedure to identify the pathway of choice of the HN–AD microbes is to verify the metabolic intermediate product formations or enzyme activity measurements. For example, in *Alcaligenes faecalis NR* and *Alcaligenes faecalis no. 4*, the accumulation of the intermediate products NO_2–N and NO_3–N was very negligible, and the corresponding enzymes were not detected.[53] This further leads to the speculation that the HN–AD process can have an alternate pathway through conversion of hydroxylamine to nitrogen directly as shown in eqn (6.13).

HN–AD Process *via* Hydroxylamine

$$NH_4^+ \xrightarrow{AMO} NH_2OH \xrightarrow{\text{Cyt P460 of HAO}} N_2O \xrightarrow{NosZ} N_2 \tag{6.13}$$

as proposed by Ren *et al.*[32]

This pathway was validated by various culture dependent analyses where the respective HN–AD species was added to the culture medium with NO_2–N and NO_3–N as the sole nitrogen source under aerobic conditions. The results indicated no removal of NO_2–N and NO_3–N.[40] But upon using hydroxylamine as the sole nitrogen source, the production of nitrogen gas was observed in the enrichment process indicating that the metabolic pathway of these HN–AD microbes is *via* hydroxylamine reduction. The hydroxylamine pathway can simplify the denitrification pathway to a larger extend improving nitrogen recycling and treatment.

6.3.1.3 *Incomplete Nitrogen Removal Pathway*

The oxidation of electron donor NH_4–N to hydroxylamine (NH_2OH) is considered as the rate limiting step in all pathways proposed for the HN–AD process. This step is mediated by the AMO enzymes common to other nitrifying microbes too. Meanwhile, aerobic denitrification is mediated by four specific reductases: (i) nitrate reductases (Nap), (ii) nitrite reductases (Nir), (iii) nitric oxide reductases (Nor) and (iv) nitrous oxide reductases (Nos). In many species specific genes coding for Nir, Nor or Nos get damaged or are absent resulting in incomplete development of denitrification capacity for a few HN–AD microbes.[51] This results in another pathway called incomplete nitrogen removal.[51] Such HN–AD microbe mediated pathways result in huge accumulation of intermediate products mainly NO_2–N/NO_3–N.[15] The incomplete pathway can happen if the required environmental conditions are altered. This effects the gene expression and enzyme activity. For example, variation in DO and low carbon source will restrain nitrite reductase leading to nitrite accumulation.[54] More research needs to be done to summarize the actual biochemical mechanisms mediated by the diverse HN–AD species.

6.3.2 Nitrogen Assimilation Mechanisms by HN–AD Microbes

One of the major attributes or versatilities of HN–AD microbes is that a few of the species have the capacity for nitrogen assimilation. Studies have reported that almost 40% of nitrogen removed can be attributed to nitrogen assimilation.[49] For instance, the denitrifying strain Y-11 was identified to use tryptone as organic nitrogen and easily converts it to biomass nitrogen and inorganic nitrogen but fails to convert it to nitrogen gas.[25] Another strain (*Pseudomonas stutzeri T13*) was found to assimilate nitrogen under aerobic and anaerobic conditions.[54]

The traditional nitrification genes (AMO and HAO genes) and denitrification genes (Nir and Nos genes) are not found in *Klebsiella* sp. Hence removal of nitrogen happens through incorporation or assimilation into cells.[55] It was reported that the assimilation happens through the NADP-glutamate dehydrogenase pathway. The pathway proposed is as shown in eqn (6.14).

Nitrogen Assimilation by HN–AD Microbes

$$NO_3^- \text{ (from organic nitrigen)} \xrightarrow{\frac{\text{Nas}}{\text{Nap}}} NO_2^- \xrightarrow{\frac{\text{NirA}}{\text{NirB}}} NH_4^+ \xrightarrow{\text{glutamate dehydrogenase}} \text{glutamate (cell growth)} \tag{6.14}$$

as proposed by Pal *et al.*[34]

Major strains classified as HN–AD microbes, identified to have the assimilation characteristics, are *Agrobacterium* sp. LAD 9, *Rhodococcus* sp. CPZ24, halophilic *Vibrio diabolicus* SF16, *Bacillus* sp. LY, *Providencia rettgeri* strain YL, *Alcaligenes faecalis* C16, *Acinetobacter calcoaceticus* HNR, and *Pseudomonas putida* strain NP5.[20] The max assimilation (in %) to biomass was reported to be 61.79% by *Pseudomonas putida* strain NP5.

Several HN–AD microbes in the Fungi phylum are found in soil. Most of the HN–AD species identified from the soil habitat were found to degrade organic carbon while a few bacteria show high assimilation capacity in an inorganic carbon environment.[34] Assimilation can improve the adaptability of the microorganisms to high concentrations of NH_4–N. The major enzymes mediating the assimilation of ammonia are (i) glutamate dehydrogenase (GDH), (ii) glutamate synthase (GS), and (iii) glutamate synthetase (GOGAT).[55]

HN–AD microbes have the capability of diverse metabolic pathways that use different organic carbon sources to either dissimilate to nitrogen gas or assimilate to organic nitrogen. In addition to this capability, recent research shows that the HN–AD microbes when in mixed culture prove more effective for the removal of nitrogen from wastewater than a single strain or pure strain. This is because of the cooperative interaction between various organisms having different nitrogen metabolic enzymes. For example, the interaction between *Pseudomonas stutzeri*, *Rhodococcus* sp. and *Bacillus subtilis* showed good removal of high concentrations of organic carbon and nitrogen from high strength wastewaters. Additionally, mixed cultures would have higher biomass production and process stability than a pure culture in the process of wastewater treatment.[56] They can withstand more changes and complexities.

6.3.3 Phosphorous Removal Pathway by HN–AD Microbes

Along with nitrogen removal, the ability to remove phosphorous proves more beneficial. It helps in abatement of the severe cause of eutrophication. The main two methods of removal of phosphorous include: (i) employing phosphorous accumulating organisms (PAOs), which uptake soluble phosphorous in the form of intracellular polyphosphate under aerobic conditions, and (ii) employing denitrifying PAOs (dPAO), which store polyphosphates under anoxic conditions. Currently, a few HN–AD species like *Pseudomonas stutzeri YG-24*, *Bacillus cereus GS-5*, *Pseudomonas putida NP5*, and *Enterobacter cloacae* are identified to have the ability of simultaneous removal of nitrogen and phosphorous from wastewaters.[30] More specific mechanics are still unclear leading to more in-depth research on the HN–AD metabolic pathway being necessary.

6.4 Factors Affecting the HN–AD Process

The applicability and performance of the HN–AD process depends on the process and environmental factors affecting growth and metabolism. Pure cultures or single strains of HN–AD microbes are used for understanding the metabolic pathway and performance evaluation with respect to the change in environmental conditions. The major factors responsible for variations in their efficiency are the concentrations and types of substrate (carbon/nitrogen sources), pH, temperature, DO requirements, inorganic ions/metal ions and salinity.[50] The diverse HN–AD species express a lot of versatility even though their mechanisms remain vague. The major factor influencing the efficiency is the availability of substrates for the HN–AD microbes.

6.4.1 Substrate Effect (Nitrogen and Carbon Source)

The basic cellular metabolic pathways of all biological water treatment technologies include carbon and nitrogen metabolism.[43] They provide the energy and assist in the improvement of cellular growth. The carbon acts as the structural unit and the nitrogen acts as the major component synthesizing amino acids, proteins, and nucleic acids.

6.4.1.1 Nitrogen Source

Nitrogen removal is the major role of HN–AD species where they utilize NH_4–N and oxidize it to NO_2–N and NO_3–N in the presence of organic carbon. The pathway followed for utilizing NH_4–N is almost the same and effective *via* AMO enzymes for all species. After the formation of NO_2–N and NO_3–N, the presence of functional genes decides the henceforth pathways as discussed in the above section. Considering the formation of NH_2OH by AMO enzymes, the nitrogen removal would be through NH_2OH as a major source. The contribution of NH_2OH to cell growth is negligible and is well noted in *Alcaligenes faecalis WT14*[36] and *Serratia marcescens W5*.[39]

The NO_2–N removal *via* denitrification was found to have a high rate in many species like *Bacillus* sp. LY,[40] and *Acinetobacter HA-2*.[29] It was found that the type of nitrogen source affected the growth of HN–AD microbes. According to Yao *et al.*,[29] the order of the growth corresponding to the nitrogen source was reported as $NH_4\text{–N} > NO_2\text{–N} > NO_3\text{–N}$. But, *Chryseobacterium* sp. *R31* was found to contradict this, as the nitrogen removal happened independent of the type of nitrogen substrate.[47]

There were reports of NO_2–N concentrations resulting in inhibition of the enzyme activity of several HN–AD species. A few studies reported that a NO_2–N concentration exceeding 50 mg L^{-1} itself was inhibitive to growth and nitrogen removal.[57] But versatile HN–AD species like *Pseudomonas tolaasii Y-11*,[25] *Pseudomonas stutzeri YZN-001*, *Alcaligenes faecalis WT14*[36] and *Acinetobacter* sp. *HA-2*[29] showed no inhibition beyond 50 mg L^{-1} of NO_2–N. *Pseudomonas tolaasii Y-11* could withstand a concentration of 200 mg L^{-1} of NO_2–N. Meanwhile, *Pseudomonas stutzeri YZN-001* and *Alcaligenes faecalis WT14* could maintain rapid growth even at NO_2–N concentrations of 100 mg L^{-1}. Also, *Acinetobacter* sp. *HA2* could withstand 80 mg L^{-1} of NO_2–N concentration. Considering NH_4–N inhibition, most of the HN–AD species are found to be effective in the removal of NH_4–N when the C/N ratio is at 10 (considered as optimum). But, the abundance of the species is found to reduce severely when the NH_4–N concentration increases above 1000 mg L^{-1}.[21] This is because of the biotoxicity caused due to high free ammonia concentrations. The nitrogen removal in real wastewater will be more complex than that in a synthetic medium due to the combination of multiple substrates. The understanding of the nitrogen metabolic activities of each HN–AD species under real scenarios is yet to be researched.

6.4.1.2 Carbon Sources

The HN–AD microbes utilize organic carbon for enhancing ammonia removal and cell growth. This is in contrast to the pathway followed by autotrophic nitrifiers as they use inorganic carbon for cellular metabolism and nutrient removal. A few HN–AD microbes like *Bacillus subtilis A1* have the capacity to mediate metabolic pathways in the presence and absence of organic carbon. The species showed successful growth in the presence of CO_2 and $NaHCO_3$ when NH_4–N was the electron donor and the sole nitrogen source. A study was conducted in the presence of COD concentration varying between 0 mg L^{-1} to 94.7 mg L^{-1}. The reaction resulted in 20.4% removal of NH_4–N in the form of nitrogen gas and lesser proportions converted to biomass and intermediate products (NO_2–N and NO_3–N).[58] This shows that *B. subtilis A1* can withstand the autotrophic environment and use the organic carbon for the denitrification process. When organic carbon in the form of acetate, glucose, citrate or succinate was provided, the species exhibited HN–AD metabolism. The performance was better using the HN–AD pathway and the performance did not differ with the types of organic carbon sources for the *B. subtilis A1* species. A similar mechanism was observed in *Serratia marcescens W5*. This species also worked in the presence and absence of organic carbon. Even though a few HN–AD species can function in the presence or the absence of organic carbon, not all HN–AD species are functional in both conditions. Zhao *et al.*[40] reported that the *Bacillus* sp. *LY* was found to exhibit nitrification only in the heterotrophic pathway. The diverse HN–AD microbes display different preferences for the organic substates.

The common organic carbon sources added for the enrichment of HN–AD species are acetate, glucose, sucrose, citrate, and succinate. The nitrification/denitrification abilities displayed by the diverse HN–AD microbes are uniquely different with respect to the carbon source preferred. Song *et al.*[20] summarizes the optimal carbon source with respect to HN–AD species. It was found that acetate was the preferred carbon source for the strains *Enterobacter cloacae HW-15*, *Bacillus subtilis A1* and *Agrobacterium* sp. *LAD9.* Citrate was identified as the desired carbon source for *Pseudomonas stutzeri YG-24* and *T1*. *Bacillus methylotrophicus strain L7* worked well in the presence of sodium succinate and glucose. In addition to biodegradable organics as substrates, toxic organic substances have also been reported to be used as a carbon source. Ge *et al.* (2015) observed that *Diaphorobacter* sp. could use phenol as the organic carbon source and simultaneously use it for nutrient removal and energy. The phenol (1400 $mg\,L^{-1}$) was completely degraded within 85 min. The NH_4–N (100%) and NO_3–N (91%) removal happened within 75 min. Liu *et al.*[59] reported the pH variation with respect to the type of organic carbon used for metabolism. For example, when the organic carbon used is pyruvate, citric acid or acetic acid, the pH was found to be higher after metabolism. On the other hand, when sucrose and glucose were used as the carbon source, the pH was decreased. pH is a very important environmental factor as it affects the growth and metabolism of HN–AD microbes.

6.4.2 Carbon to Nitrogen Ratio (C/N Ratio)

This is an important and vital parameter for understanding the microbial cell growth and functioning. The type of treatment processes or type of microbial interactions would be decided by this parameter. As the autotrophs generally function at a C/N ratio ≤ 1, a high concentration of organic carbon would be harmful for the functioning of autotrophic nitrifiers. Hence, the traditional BNR system would be functioning effectively after reducing the C/N ratio.[20]

Contrary to traditional BNR, the HN–AD microbes require a higher C/N and an optimum C/N can improve the nitrification rate. HN–AD species have a wide adaptation range for C/N ratio and the optimum is mostly reported at C/N = 10. The feasible range of C/N ratio for many HN–AD species was reported to be between 2 and 15,[15,20] where most of the species exhibit the optimum ratio between 8 and 12. Higher C/N ratios are more beneficial for the HN–AD species as they can accelerate the growth of heterotrophic nitrifiers while lower C/N ratio affects the denitrification capacity.[23,43]

There are observations that the removal is affected when the C/N ratio is deviated to lower or higher than the optimum range. When the C/N ratio is lower (C/N $\leq$ 5), the HN–AD reactions get impaired due to lack of sufficient carbon to complete denitrification. Hence, the need for an external supply of carbon would be there, which is cost inclusive. When the C/N ratio is higher, *i.e.*, C/N = 10, the denitrification rate increases 5 to 40 times higher than for other bacteria that can denitrify. When the C/N ratio is more than optimum, the nitrogen removal is found to improve, while the COD removal gets affected. Xi *et al.*[15] reported efficient nitrogen and COD removal by *Exiguobacterium mexicanum SND-01* at a C/N ratio of 12. The species *Acinetobacter baumannii AL-6* was found to effectively perform HN–AD activities within the optimum range of 10–15. When the C/N ratio is less than 10, the efficiency of HN–AD can be improved by adding external carbon sources. Such observations were reported by Silva *et al.*[27] Significant nitrogen removal was observed at the C/N ratio of 8 by *Pseudomonas balearica UFV3* and *Gordonia amicalis UFV4*. *Enterobacter cloacae HW-15* also exhibited similar performance at a C/N ratio of 8.[22] Many strains exhibit optimum removal of NH_4–N at the C/N ratio of 10, but even at this ratio the chances of inhibition were inevitable when the concentrations of NH_4–N were high.[15] Comparing all HN–AD species, the various species of *Bacillus* Genera can tolerate a wide range of C/N ratios (2 to 20). The HN–AD species were found to easily adapt to mixed culture consortiums also, which makes the competition more complex. The C/N ratio is the major factor responsible for the combined reaction or competition of one species over the other.

6.4.3 Environmental Factors (pH and Temperature)

The environmental factors are the important guiding parameters for a biochemical reaction. Most of the species are sensitive to any changes to these parameters. These factors influence the enzyme activity and concentrations of free ammonia (FA).[60]

6.4.3.1 Optimum pH for HN–AD Microbes

The effectiveness of nutrient removal processes is highly dependent on the pH. The conventional nitrification process effectively works under weakly alkaline conditions and a drop in pH towards the acidic range can inhibit the nitrification process. This is mainly because the acidic pH increases the accumulation of free ammonia (FA) in the system. This also affects the enzymatic activity of AMOs. Compared with autotrophic nitrifies, the HN–AD microbes are considered to have better acid tolerance. HN–AD microbes classified under the phylum Fungi have been reported to have better acid-resisting capability. They oxidize organic nitrogen and ammonium in the soil.[56] There are a few acidophilic bacterial species also classified as HN–AD microbes.[15] *Penicillium* sp. *M25-22* was reported to tolerate a pH range of 5 to 8.5. But the optimum pH was found at 7.5. Species such as *Stenotrophomonas* sp. *MSNA-1 Acinetobacter* sp. *JR1* and *Acinetobacter junii YB* were found to be acid tolerant and were observed to have pH tolerance ranging from 3 to 11. The HN–AD microbes are reported to have a wide range in pH adaptability. Also, the acid tolerance can be improved if the selected strain is made to enrich in a low pH environment.[42] This makes them suitable to be employed for treating acid wastewaters. The optimal pH range for most of the HN–AD microbes is found to range between 6 and 9 as summarized by Xi *et al.*[15]

Generally, the HN–AD microbes grow better in alkaline conditions. The species *Bacillus litoralis N31* shows efficient nitrification rates in weakly alkaline conditions (7.5 to 8.5).[16] *Halomonas* sp. *BO1* was able to efficiently remove nitrogen at pH ranging between 8 and 9. *Pseudomonas balearica UFV3* also preferred a pH ranging between 7 and 9 for complete removal of nitrogen. Most of the species were seen to exhibit an optimum between 7.5 to 8.5 or maximum 9. Apparently the critical pH point of acid and alkaline resistant species to exhibit nitrogen conversion ability was reported as 10. More than pH 10, the formation of FA happens and would result in inhibiting the nitrification activity.[15]

6.4.3.2 Optimum Temperature for HN–AD Microbes

Most of the nitrogen conversions reported by HN–AD species are sensitive to temperature variations. Generally, all HN–AD microbes are mesophylls. Their activity is optimum in the temperature range 20 to 40 °C (as summarized in ref. 20). Even though small temperature changes bring about a significant effect in the activity of HN–AD microbes, several species have shown versatility in undergoing the nitrification and denitrification reactions under low and high temperatures. Table 6.1 shows many psychrophilic and a single thermophilic HN–AD species which removed nitrogen with high efficiency. The adaptation exhibited by the HN–AD species to varying temperatures reveals the great potential for employing the HN–AD process for abatement of nitrogen from wastewaters in the cold region.

6.4.4 Dissolved Oxygen Requirements and Tolerance

Dissolved oxygen (DO) is an important parameter that has the capacity to regulate the rate of reaction. The availability of DO decides the efficiency of removal of NH_4–N, NO_2–N and NO_3–N from the system when employing HN–AD microbes. The HN-A process is identified as an aerobic process where nitrification and denitrification require oxygen. This can be an anticipated disadvantage due to the extensive energy requirement associated with the working of the process. But, other advantages of the process such as: (i) high yield, (ii) efficient removal at high C/N ratio and high concentration, and (iii) treatment being possible in a single reactor, *etc.* make the process preferable.

The DO regulates various reactions in HN–AD microbes by mediating enzymatic activities related to nitrogen conversions.[34] In most of the studies, the DO concentration in the reactors is maintained by constant shaking at varying speeds. It is reported that the shaking speeds of 120, 150 and 160 rpm were equated to give a DO concentration of 4.7, 6.5 and 7.2 mg L^{-1} respectively.[15] Hence, a similar approach was considered for other studies too. Sun *et al.*,[54] reported a high NO_2–N removal when the shaking was maintained at 160 rpm. The *Halomonas BO1* strain was initially operated without shaking resulting in a low (41.5%) removal efficiency. When the shaking was increased to 90 rpm, the removal increased to 96%.[23] DO is required at high concentrations for heterotrophic nitrification. Considering major HN–AD species, the optimum range for DO was from 3 to 6.75 mg L^{-1}. *Pseudomonas species* were found to work well in low DO compared to other HN–AD species. *Pseudomonas AND-42* performed efficient nitrogen removal at a DO concentration of 3 mg L^{-1}.[28] *Marinobacter* sp. *F6* was found to require a high DO of 6.75 mg L^{-1} for efficient removal.[24]

The major change from the traditional BNR process in the HN–AD process is the possibility of aerobic denitrification in HN–AD microbes. Different HN–AD microbes are likely to have different optimal DO conditions for undergoing denitrification. It is reported that[17,49] species such as *Alcaligenes faecalis no. 14*, *Acinetobacter* sp. *ND7* and *Cupriavidus* sp. *S1* have a profound adaptation ability to DO. The denitrification capacity is found to decrease gradually with increase in the DO concentration in HN–AD species. The species that could undergo complete denitrification in the presence of increased DO were *Alcaligenes faecalis TUD* and *Pseudomonas stutzeri YG-24.*[23] The regulation of the optimum DO to enable heterotrophic nitrification and denitrification will improve the efficiency.

6.4.5 Effect of Inorganic Ions on HN–AD Microbes

Many inorganic ions were reported to have an influence on the growth and enhancement of HN–AD microbes.[20] Metals like iron (Fe^{2+} and Fe^{3+}), calcium, zinc, magnesium and manganese were reported to show enhanced nitrogen removal when present with HN–AD microbes. Some ions can increase the enzyme activity, while a few are rendered as inhibitors also. The ions Cu^{2+}, Zn^{2+}, Ni^{2+} and Cr^{6+} were reported to inhibit strains like *Cupriavidus* sp. *S1* and *Pseudomonas putida ZN1.*[16,20] The enzyme activity of AMO was found to be enhanced by the addition of Mg^{2+}, Zn^{2+} and Mn^{2+}.[23] More studies are required in this area for understanding the role played by the inorganic ions in detail.

6.4.6 Salinity Effect on HN–AD Microbes

Salinity is a major inhibitive factor for microbes as high concentrations of salinity result in cell lysis due to an increase in osmotic pressure, a change in microbial metabolism and inhibition of enzyme activity.[51] The salt reduces the transport capability within the cells. It also leads to dehydration and cell lysis.[53]

It was observed that NOBs tolerate a low concentration of NaCl (20 g L^{-1}),[61] while AOBs could tolerate about 30 g L^{-1} of salinity.[24] Hence their applicability in treating saline wastewater is questionable. A few HN–AD strains were discovered to be halophilic, such as *Aeromonas* sp. *HN-02, Bacillus MS 30*, *Bacillus methylotrophicus strain L7*, *Pseudomonas*

TJPUo4 and *halophilic Vibrio diabolicus SF16.*[20] Such strains can play a major role in the abatement of nitrogen from salt-containing wastewater. With this diverse adaptability the applicability of HN–AD microbes is extensive.

6.5 Application of the HN–AD Process in Wastewater Treatment Facilities

The diversity of HN–AD microbes is very extensive, such that employing the HN–AD process to treat and improve the existing remediation potential is feasible. The initial research was more focused on isolation and identification of pure strains of HN–AD microbes. Later, their application to treat high strength wastewaters was verified under controlled conditions. Recently, more emphasis has been given to the incorporation of mixed cultures for treating complex wastewaters or multipollutant removal processes. Table 6.2 summarizes recent studies done using the HN–AD process towards the abatement of nitrogen from wastewaters. The major results comparing these studies show that the application of HN–AD microbes in real wastewater produces promising results, but these results are mostly obtained from lab scale or pilot scale reactors. Until now, only one field scale reactor has been developed for the application of HN–AD microbes. Chen *et al.*[31] reported an anaerobic/anoxic/aerobic field scale reactor for treating piggery wastewater. The HN–AD strain used was *Acinetobacter* sp. *T1.* The pure culture was directly bioaugmented in aeration tanks of 300 m^3 capacity. The results after 28 days of investigation showed an overall enhancement of NH_4-N, TN, and COD removal. Acclimatized cultures can also be added to the reactors for enhancing the removal. By acclimatizing, the HN–AD species becomes more adapting to variations in C/N ratios.

The major constraint is to enrich and maintain a pure culture on a large scale for treating real wastewaters.[21] Other constraints include: (i) difficulty in improving the removal efficiency using the one stage HN–AD process alone. Researchers have identified that when multiple processes are combined or different microorganisms mediating different processes coexist, it can improve the overall efficiency. For example, wastewater containing high strength organic carbon can be pretreated using anaerobic digestion succeeded by the HN–AD process. In this, the HN–AD process can effectively remove the remaining organic carbon and nitrogen compounds bringing an overall abatement of C and N pollution.

(ii) Studies are mainly designed for treating synthetic wastewater; the robustness of the process needs to be verified by employing it in real wastewaters. The complexities would be more and hence the adaptability of the HN–AD process can be verified. (iii) There are lab scale reactors involving pure and mixed culture species, but it lacks in-depth analysis giving information about interspecific interactions including mutualism, predation, and interspecies competition. This will help to elucidate more in-depth knowledge to understand the underlying mechanisms in operation of

Table 6.2 Recent studies on the HN–AD process employed for nitrogen removal.

Sl. no.	Type of culture	Strain used	Type of wastewater treated	Type of reactor	Performance	Ref.
1.	Pure culture	*Pseudomonas* sp. and *Rhodoferax ferrireducens*	Synthetic wastewater	SBR (lab scale)	A psychrotolerant consortium was developed with excellent nitrogen removal capability (specific ammonia and nitrate removal rate being 3.1 mg N per (g SS h) and 9.6 mg N per (g SS h)	29
2.	Pure culture	*Diaphorobacter* sp. *PD-7*	Synthetic wastewater containing phenol	Batch reactor	Diaphorobacter sp. PD-7 could remove ammonium and nitrate at a rate of 1.61 and 1.16 $mg\,L^{-1}\,h^{-1}$ and 100% phenol was degraded within 85 h	37
3.	Pure culture	*Vibrio diabolicus SF16.*	Saline wastewater	Aerated biological filter system	91.82% of NH_4–N (119.77 $mg\,L^{-1}$) and 99.71% of NO_3–N (136.43 $mg\,L^{-1}$). The nitrogen balance showed that 35.83% of initial NH_4–N (119.77 $mg\,L^{-1}$) was changed to intracellular nitrogen, and 53.98% of the initial NH_4–N was converted to gaseous denitrification products.	51
4.	Pure culture	*Bacillus Licheniformis*	Synthetic aquarium wastewater	Aquarium + porous material (attached growth system, lab scale)	The efficiencies of NH_4–N and TN removal reached 65–66% at NH_4–N of 20 and 30 $mg\,L^{-1}$	44

5	Pure culture	*Acinetobacter* sp. *T1*	Piggery wastewater	Anaerobic/anoxic/aerobic reactor (full scale)	COD, NH_4–N and TN removal efficiencies were enhanced by 30%, 15% and 16%	31
6.	Pure culture	*Stenotrophomonas maltophilia DQ01*	Synthetic wastewater (varying C/N ratio)	Moving bed biofilm reactor	Efficiency (94.21%) and total nitrogen removal (94.43%) were achieved at C/N of 7.5, and declined with C/N decreasing due to the reduction of electron donation and consumption.	13
7.	Mixed culture	*Heterotrophic nitrifying Stenotrophomonas* sp. *MSNA-1* and *aerobic denitrifying Pseudomonas* sp. *MSD*	Wastewater from chemical plant	Lab scale SBR reactor	Both cultures exhibited extreme tolerance to salinity and pH without affecting nitrogen removal	42
8.	Mixed culture	*Acinetobacter, Pseudomonas* and *Paracoccus, other bacteria Halomonas* and *Thauera*	High ammonia wastewater	Moving bed biofilm reactor (lab scale)	Start-up time was shortened to 15 days and NH_4^+ tolerance was enhanced to 1000 mg L^{-1}	62
9.	Mixed culture	*Paracoccus (55%) dominated HN–AD consortium*	Synthetic saline wastewater	Halophilic aerobic granular sludge system (HAGS) (lab scale)	100% TIN removal happened aerobically	63

HN–AD microbes. Furthermore, this will help in providing more understanding for employing the process for multi pollutant abatement from real wastewaters.

6.5.1 Future Scope of Research in the HN–AD Process

The HN–AD process is a promising process with a lot of versatility in terms of tolerance to adverse environmental conditions, diverse habitat, easy start up and high yield of microbes, attaining fast process stability, *etc.* Therefore, there exists a lot of scope in employing HN–AD species to solve the organic carbon and nutrient pollution in wastewaters. A few of the areas where further research is required are listed below.

(i) In-depth research into the general molecular information of HN–AD species *via* modern omics approaches (genomics, transcriptomics, and proteomics).
(ii) Research on the inclusion of more mixed culture processes and understanding of the characteristics exhibited by HN–AD microbes in such a microbial consortium.
(iii) The major nitrogen removal pathways are hypothesized but the research on key enzymes and pathways is inadequate. A detailed study of the biochemical reactions can provide an appropriate solution to efficient nutrient removal using the HN–AD process.
(iv) Research on full scale reactors or practical integrated systems employing the HN–AD process is needed.
(v) A proper characterization or enrichment technique still needs to be envisaged.
(vi) Profound analysis is required to study the inhibitory effects on HN–AD microbes.
(vii) Not much information is available for fungal species in HN–AD. Fungal species are highly diverse and compared with bacteria, filamentous fungi have the advantage of being easy to harvest due to the mycelium growth and greater resistance to toxic and inhibitory compounds.[64] Hence such studies would be more beneficial regarding the application of the HN–AD process for micro-pollutant removal.

6.6 Conclusions

Heterotrophic Nitrification and Aerobic Denitrification (HN–AD) is a contrary process to conventional nitrification and denitrification processes. In this process, the reactive nitrogen species (NH_4–N) is converted to NO_2–N/NO_3–N and then to N_2 in the presence of organic carbon and oxygen. The HN–AD microbes follow multiple pathways based on the functional genes/enzymes present in them. The special microbes mediating this process fall under 6 phylums, namely Proteobacteria (gamma, beta, alpha),

Firmicutes, Fungi, Bacteroidota, Actinobacteria and Halomonadaceae. There are almost 58 species reported to date classified into these phyla. All species are from diverse habitats like wastewater treatment plants, marine sediments, landfill leachates, activated sludge, piggery or swine wastewater, soil, sewage, deep sea hydrothermal vents, surface water, *etc.* This diversity brings in high adaptability as their virtue. The major factors for the optimum growth have been identified as: (i) optimum pH 7.5 to 8.5, a few species have acid tolerance capacity, (ii) optimum temperature: mesophilic range (20 to 35 °C), with a few species as psychrotolerant, and (iii) optimum C/N ratio varied between 2 and 15, with favorable range 8 to 12.

A few HN–AD species are found to withstand salinity to a max of 30 $g L^{-1}$. Being aerobic microbes, they require sufficient oxygen (min $\geq$ 2 $mg L^{-1}$) for proceeding with the HN–AD reaction. The reactors employing HN–AD have mostly been utilized for removing pollutants from synthetically prepared wastewater. The requirement of full-scale plants is necessary to improve the application of the HN–AD process for treating complex wastewaters.

References

1. B. Cui, Q. Yang, Y. Zhang, X. Liu, W. Wu and J. Li, Improving nitrogen removal in biological aeration filter for domestic sewage treatment *via* adjusting microbial community structure, *Bioresour. Technol.*, 2019, **293**, 122006.
2. J. Li, Q. Du, H. Peng, Y. Zhang, Y. Bi, Y. Shi, Y. Xu and T. Liu, Optimization of biochemical oxygen demand to total nitrogen ratio for treating landfill leachate in a single-stage partial nitrification-denitrification system, *J. Cleaner Prod.*, 2020, **266**, 121809.
3. Y. Feng, Y. Peng, B. Wang, B. Liu and X. Li, A continuous plug-flow anaerobic/aerobic/anoxic/aerobic (AOAO) process treating low COD/TIN domestic sewage: Realization of partial nitrification and extremely advanced nitrogen removal, *Sci. Total Environ.*, 2021, **771**, 145387.
4. W. Gujer, 'Nitrification and me? A subjective review', *Water Res.*, 2010, **44**(1), 1–19.
5. S. F. Kosari, B. Rezania, K. V. Lo and D. S. Mavinic, 'Operational strategy for nitrogen removal from centrate in a two-stage partial nitrification-anammox process', *Environ. Technol.*, 2014, **35**, 1110–1120.
6. Y.-H. Ahn, Sustainable nitrogen elimination biotechnologies: A review, *Process Biochem.*, 2006, **206**(41), 1709–1721.
7. G. M. Islam, P. Vi and K. A. Gilbride, Functional relationship between ammonia-oxidizing bacteria and ammonia-oxidizing archaea populations in the secondary treatment system of a full-scale municipal wastewater treatment plant, *J. Environ. Sci.*, 2019, **86**, 120–130.
8. I. Cotto, Z. Dai, L. Huo, C. L. Anderson, K. J. Vilardi, U. Ijaz, W. Khunjar, C. Wilson, H. De Clippeleir, K. Gilmore and E. Bailey, Long solids retention times and attached growth phase favor prevalence of comammox bacteria in nitrogen removal systems, *Water Res.*, 2020, **169**, 115268.

9. S. Yin, J. Li, H. Dong and Z. Qiang, Enhanced nitrogen removal through marine anammox bacteria (MAB) treating nitrogen-rich saline wastewater with Fe(III) addition: Nitrogen shock loading and community structure, *Bioresour. Technol.*, 2019, **287**, 121405.
10. X. X. Wang, S. Y. Wang and T. L. Xue, *et al.*, Treating low carbon/nitrogen (C/N) waste-water in simultaneous nitrification-endogenous denitrification and phosphorous removal (SNDPR) systems by strengthening anaerobic intracellular carbon storage, *Water Res.*, 2015, **77**, 191–200.
11. Q. Chen and J. Ni, Ammonium removal by agrobacterium sp. LAD9 capable of heterotrophic nitrification–aerobic denitrification, *J. Biosci. Bioeng.*, 2012, **113**(5), 619–623.
12. X. Huang, W. Li, D. Zhang and W. Qin, Ammonium removal by a novel oligotrophic acinetobacter sp. Y16 capable of heterotrophic nitrification–aerobic denitrification at low temperature, *Bioresour. Technol.*, 2013, **146**, 44–50.
13. Y. Jia, M. Zhou, Y. Chen, Y. Hu and J. Luo, Insight into short-cut of simultaneous nitrification and denitrification T process in moving bed biofilm reactor: Effects of carbon to nitrogen ratio, *Chem. Eng. J.*, 2020, **400**, 125905.
14. L. A. Robertson, E. W. Niel Van and R. A. Torremans, Simultaneous nitrification and denitrification in aerobic chemostat cultures of Thiosphaera pantotropha, *Appl. Environ. Microbiol.*, 1988, **54**, 2812–2818.
15. H. Xi, X. Zhou, M. Arslan, Z. Luo, J. Wei, Z. Wu and M. G. El-Din, Heterotrophic nitrification and aerobic denitrification process: Promising but a long way to go in the wastewater treatment, *Sci. Total Environ.*, 2022, **805**, 150212.
16. F. Huang, L. Pan, N. Lv and X. Tang, Characterization of novel bacillus strain N31 from mariculture water capable of halophilic heterotrophic nitrification–aerobic denitrification, *J. Biosci. Bioeng.*, 2017, **124**, 564–571.
17. Z. Sun, Y. Lv and Y. Liu, *et al.*, Removal of nitrogen by heterotrophic nitrification–aerobic denitrification of a novel metal resistant bacterium cupriavidus sp. S1, *Bioresour. Technol.*, 2016, **220**, 142–150.
18. H. Wang, J. Li and B. Wang, *et al.*, Deciphering pollutants removal mechanisms and genetic responses to ampicillin stress in simultaneous heterotrophic nitrification and aerobic denitrification (SHNAD) process treating seawater-based wastewater, *Bioresour. Technol.*, 2020, **315**, 123827.
19. F. Huang, L. Pan, Z. He, M. Zhang and M. Zhang, Culturable heterotrophic nitrification–aerobic denitrification bacterial consortia with cooperative interactions for removing ammonia and nitrite nitrogen in mariculture effluents, *Aquaculture*, 2020, **523**, 735211.
20. T. Song, X. Zhang, J. Li, X. Wu, H. Feng and W. Dong, A review of research progress of heterotrophic nitrification and aerobic denitrification microorganisms (HNADMs), *Sci. Total Environ.*, 2021, **801**, 149319.

21. M. Zhang, A. Li, Q. Yao, Q. Wu and H. Zhu, Nitrogen removal characteristics of a versatile heterotrophic nitrifying-aerobic denitrifying bacterium, *Pseudomonas bauzanensis DN13-1*, isolated from deep-sea sediment, *Bioresour. Technol.*, 2020, **305**, 122626.
22. W. J. Wan, D. L. He and Z. J. Xue, Removal of nitrogen and phosphorus by heterotrophic nitrification–aerobic denitrification of a denitrifying phosphorus-accumulating bacterium Enterobacter cloacae HW-15, *Ecol. Eng.*, 2017, **99**, 199–208.
23. C. Li, J. Yang, X. Wang, E. Wang, B. Li, R. He and H. Yuan, Removal of nitrogen by heterotrophic nitrification–aerobic denitrification of a phosphate accumulating bacterium Pseudomonas stutzeri YG-24, *Bioresour. Technol.*, 2015, **182**, 18–25.
24. Q. Zhang, Y. Liu, G. Ai, L. Miao, H. Zheng and Z. Liu, The characteristics of a novel heterotrophic nitrification–aerobic denitrification bacterium, *Bacillus methylotrophicus strain L7*, *Bioresour. Technol.*, 2012, **108**, 35–44.
25. T. He, Z. Li, Q. Sun, Y. Xu and Q. Ye, Heterotrophic nitrification and aerobic denitrification by *Pseudomonas tolaasii Y-11* without nitrite accumulation during nitrogen conversion, *Bioresour. Technol.*, 2016, **200**, 493–499.
26. F. Xie, M. Thiri and H. Wang, Simultaneous heterotrophic nitrification and aerobic denitrification by a novel isolated *Pseudomonas mendocina X49*, *Bioresour. Technol.*, 2021, **319**, 124198.
27. L. C. F. Silva, H. S. Lima and T. A. D. Mendes, *et al.*, Heterotrophic nitrifying/aerobic denitrifying bacteria: ammonium removal under differentphysical-chemical conditions and molecular characterization, *J. Environ. Manage.*, 2019, **248**, 109294.
28. R. Jin, T. Liu and G. Liu, *et al.*, Simultaneous heterotrophic nitrification and aerobic denitrification by the marine origin *bacterium pseudomonas* sp. *ADN-42*, *Appl. Biochem. Biotechnol.*, 2015, **175**, 2000–2011.
29. S. Yao, J. R. Ni and T. Ma, *et al.*, Heterotrophic nitrification and aerobic denitrification at low temperature by a newly isolated bacterium, *Acinetobacter* sp. *HA2*, *Bioresour. Technol.*, 2013, **139**, 80–86.
30. J. R. Yang, Y. Wang and H. Chen, *et al.*, Ammonium removal characteristics of an acid-resistant bacterium *Acinetobacter* sp. *JR1* from pharmaceutical wastewater capable of heterotrophic nitrification–aerobic denitrification, *Bioresour. Technol.*, 2019, **274**, 56–64.
31. S. Chen, S. He and C. Wu, *et al.*, Characteristics of heterotrophic nitrification and aerobic denitrification bacterium *Acinetobacter* sp. *T1* and its application for pig farm wastewater treatment, *J. Biosci. Bioeng.*, 2019, **127**, 201–205.
32. Y. X. Ren, L. Yang and X. Liang, The characteristics of a novel heterotrophic nitrifying and aerobic denitrifying bacterium, *Acinetobacter junii YB*, *Bioresour. Technol.*, 2014, **171**, 1–9.
33. M. Wang, G. Cao, N. Feng and Y. Pan, Bioaugmentation of two-stage aerobic sequencing batch reactor with mixed strains for high nitrate nitrogen wastewater treatment, *Chin. J. Chem. Eng.*, 2020, **28**, 3103–3109.

34. R. R. Pal, A. A. Khardenavis and H. J. Purohit, Identification and monitoring of nitrification and denitrification genes in *Klebsiella pneumoniae EGD-HP19-C* for its ability to perform heterotrophic nitrification and aerobic denitrification, *Funct. Integr. Genomics*, 2015, **15**, 63–76.
35. Z. Wang, C. Chen, H. Liu, D. Hrynshpan, T. Savitskaya, J. Chen and J. Chen, Enhanced denitrification performance of *Alcaligenes* sp. *TB* by Pd stimulating to produce membrane adaptation mechanism coupled with nanoscale zero-valent iron, *Sci. Total Environ.*, 2020, **708**, 135063.
36. H. Chen, W. Zhou, S. Zhu, F. Liu, L. Qin, C. Xu and Z. Wang, Biological nitrogen and phosphorus removal by a phosphorus-accumulating bacteria *Acinetobacter* sp. *strain* C-13 with the ability of heterotrophic nitrification–aerobic denitrification, *Bioresour. Technol.*, 2021, **322**, 124507.
37. Q. L. Ge, X. P. Yue and G. Y. Wang, Simultaneous heterotrophic nitrification and aerobic denitrification at high initial phenol concentration by isolated bacterium *diaphorobacter* sp. *PD-7*, *Chin. J. Chem. Eng.*, 2015, **23**, 835–841.
38. J. Ren, X. Bai, Y. Liu and X. Huang, Simultaneous nitrification and aerobic denitrification by a novel isolated *Ochrobactrum anthropi HND19*, *Bioresour. Technol.*, 2021, **340**, 125582.
39. T. Wang, Q. Dang and C. Liu, *et al.*, Heterotrophic nitrogen removal by a newly-isolated alkali tolerant microorganism, *Serratia marcescens W5*, *Bioresour. Technol.*, 2016, **211**, 618–627.
40. B. Zhao, Y. L. He and X. F. Zhang, Nitrogen removal capability through simultaneous heterotrophic nitrification and aerobic denitrification by *bacillus* sp. *LY*, *Environ. Technol.*, 2010, **31**, 409–416.
41. P. Chen, J. Li and Q. X. Li, *et al.*, Simultaneous heterotrophic nitrification and aerobic denitrification by bacterium *Rhodococcus* sp. *CPZ24*, *Bioresour. Technol.*, 2012, **116**, 266–270.
42. X. Zeng, J. J. Huang, B. Hua and P. Champagne, Nitrogen removal bacterial strains, MSNA-1 and MSD4, with wide ranges of salinity and pH resistances, *Bioresour. Technol.*, 2020, **310**, 123309.
43. Q. Chen and J. Ni, Heterotrophic nitrification–aerobic denitrification by novel isolated bacteria, *J. Ind. Microbiol. Biotechnol.*, 2011, **38**, 1305–1310.
44. W. Khanichaidecha, A. Nakaruk and K. Ratananikom, *et al.*, Heterotrophic nitrification and aerobic denitrification using pure-culture bacteria for wastewater treatment, *J. Water Reuse Desalin.*, 2019, **9**, 10–17.
45. B. Dalecka, M. Strods, T. Juhna and G. K. Rajarao, Removal of total phosphorus, ammonia nitrogen and organic carbon from non-sterile municipal wastewater with *Trametes versicolor* and *Aspergillus luchuensis*, *Microbiol. Res.*, 2020, **241**, 126586.
46. J. Fang, S. Lia, S. Zhang, L. Li, S. Tan, W. Li, A. Wang and J. Ye, Characteristics of a novel heterotrophic nitrification–aerobic denitrification yeast, Barnettozyma californica K1, *Bioresour. Technol.*, 2021, **339**, 125665.

47. P. Kundu, A. Pramanik, A. Dasgupta, S. Mukherjee and J. Mukherjee, Simultaneous heterotrophic nitrification and aerobic denitrification by *Chryseobacterium* sp. *R31* isolated from abattoir wastewater, *Biomed. Res. Int.*, 2014, 436056.
48. J. Wehrfritz, A. Reilly, S. Spiro and D. J. Richardson, Purification of hydroxylamine oxidase from *thiosphaera pantotropha*: identification of electron acceptors that couple heterotrophic nitrification to aerobic denitrification, *FEBS Lett.*, 1993, **335**, 246–250.
49. L. Xie, X. Li and W. Fan, *et al.*, Heterotrophic nitrification and aerobic denitrification by a novel *Acinetobacter* sp. *ND7* isolated from municipal activated sludge, *Bioresour. Technol.*, 2020, **301**, 122749.
50. Z. Zhu, Y. Yang, A. Fang, Y. Lou, G. Xie, N. Ren and D. Xing, Quorum sensing systems regulate heterotrophic nitrification–aerobic denitrification by changing the activity of nitrogen-cycling enzymes, *Environ. Sci. Ecotechnol.*, 2020, **2**, 100026.
51. J. Duan, H. Fang, B. Su, J. Chen and J. Lin, Characterization of a halophilic heterotrophic nitrification–aerobic denitrification bacterium and its application on treatment of saline wastewater, *Bioresour. Technol.*, 2015, **179**, 421–428.
52. S. Tsujino, C. Uematsu, H. Dohra and T. Fujiwara, Pyruvic oxime dioxygenase from heterotrophic nitrifier Alcaligenes faecalis is a non-heme Fe(II)-dependent enzyme homologous to class II aldolase, *Sci. Rep.*, 2017, **7**, 39991.
53. M. Shoda, Heterotrophic nitrification and aerobic denitrification by *Alcaligenes faecalis No. 4*, *Nitrification and Denitrification*, 2017, p. 31.
54. Y. Sun, A. Li, X. Zhang and F. Ma, Regulation of dissolved oxygen from accumulated nitrite during the heterotrophic nitrification and Aerobic denitrification of *Pseudomonas stutzeri T13*, *Appl. Microbiol. Biotechnol.*, 2015, **99**, 3243–3248.
55. P. Jin, Y. Chen, R. Yao, Z. Zheng and Q. Du, New insight into the nitrogen metabolism of simultaneous heterotrophic nitrification–aerobic denitrification bacterium in mRNA expression, *J. Hazard. Mater.*, 2019, **371**, 295–303.
56. Y. Yang, Y. Liu, Y. Yang and Y. Lv, Characterization of a microbial consortium capable of heterotrophic nitrifying under wide C/N range and its potential application in phenolic and coking wastewater, *Biochem. Eng. J.*, 2017, **120**, 33–40.
57. C. Wan, X. Yang, D. Lee, M. Du, F. Wan and C. Chen, Aerobic denitrification by novel isolated strain using NO_2–N as nitrogen source, *Bioresour. Technol.*, 2011, **102**, 7244–7248.
58. X. Yang, S. Wang, D. Zhang and L. Zhou, Isolation and nitrogen removal characteristics of an aerobic heterotrophic nitrifying–denitrifying bacterium, *Bacillus subtilis A1*, *Bioresour. Technol.*, 2011, **102**, 854–862.
59. Y. Liu, Y. Wang, Y. Li, H. An and Y. Lv, Nitrogen removal characteristics of heterotrophic nitrification–aerobic denitrification by *Alcaligenes faecalis C16*, *Chin. J. Chem. Eng.*, 2015, **23**, 827–834.

60. J. Su, M. Ma and F. Ma, *et al.*, Kinetic analysis of heterotrophic nitrification–aerobic denitrification by an oligotrophic *Acinetobacter* sp. *SYF26*, *Environ. Eng. Sci.*, 2017, **34**, 844–851.
61. Z. Pan, J. Zhou and Z. Lin, *et al.*, Effects of COD/TN ratio on nitrogen removal efficiency, microbial community for high saline wastewater treatment based on heterotrophic nitrification–aerobic denitrification process, *Bioresour. Technol.*, 2020, **301**, 122726.
62. Q. Zhang, X. Chen, Z. Zhang, W. Luo, H. Wu, L. Zhang, X. Zhang and T. Zhao, Performance and microbial ecology of a novel moving bed biofilm reactor process inoculated with heterotrophic nitrification–aerobic denitrification bacteria for high ammonia nitrogen wastewater treatment, *Bioresour. Technol.*, 2020, **315**, 123813.
63. J. L. Huang, Y.-W. Cui, J. L. Yan and Y. Cui, Occurrence of heterotrophic nitrification–aerobic denitrification induced by decreasing salinity in a halophilic AGS SBR treating hypersaline wastewater, *Chem. Eng. J.*, 2022, **431**, 134133.
64. Y. Ye, J. Gan and B. Hu, Screening of phosphorus-accumulating Fungi and their potential for phosphorus removal from waste streams, *Appl. Biochem. Biotechnol.*, 2015, **177**, 1127–1136.

CHAPTER 7

Ecophysiology and Genomics of N-cycling Microbes in the Environment

JYOTI K. CHOUDHARI,[a] J. ANANDKUMAR,[b] MUKESH K. VERMA,[b,d] JYOTSNA CHOUBEY[c] AND BIJU P. SAHARIAH,*[d]

[a] Maulana Azad National Institute of Technology, Bhopal-462003, Madhya Pradesh, India; [b] National Institute of Technology Raipur, Raipur - 492010, Chhattisgarh, India; [c] Raipur Institute of Technology Raipur, Raipur - 492001, Chhattisgarh, India; [d] Chhattisgarh Swami Vivekanand Technical University, Bhilai-491107, Chhattisgarh, India
*Email: biju.sahariah@gmail.com

7.1 Nitrogen in the Biosphere and Nitrogen Cycle

Nitrogen (N_2) is the most abundant element making up 79% of the Earth's atmosphere. It can exist in 9 oxidation states, which exchange electrons (donate/accept) for electron flow as well as energy conservation. These include molecular nitrogen (N_2, oxidation state: 0), organic nitrogen, which is basically in reduced forms (for example, urea, amines, proteins, nucleic acids, *etc.*), inorganic forms of reduced N [for example, NH_3, NH_4^+ (oxidation state: −3), N_2H_4 (oxidation state: −2), and NH_4OH (oxidation state: −1)], inorganic oxidized N [for example, N_2O (oxidation state: in +1), NO (oxidation state: in +2), HNO_2 & NO_2^- (oxidation state: in +3), NO_2 & NO_3^- (oxidation state: in +4), and NO_3 & HNO_3 (oxidation state: +5)], *etc.* Nitrogen is one of the principal constituents for biosynthesizing nucleic acids, proteins and many other important biomolecules in living species. However, similar to the water content

Ammonia Oxidizing Bacteria: Applications in Industrial Wastewater Treatment
Edited by Maulin P. Shah

Published by the Royal Society of Chemistry, www.rsc.org

on the Earth, where there is only a restricted portion of bioavailable water, the bioavailable nitrogen content is also very limited. Due to the extreme involvement of nitrogen in the biosphere/biomolecules, it has attracted the attention of scientists since the early nineteenth century, and extensive studies have been conducted on the nitrogen cycle that cover the transformation of nitrogen from its gaseous form to various bioavailable forms and then its recycling to its original form. Nitrogen in its atmospheric form is not applicable for utilization in biomolecule synthesis and not all organisms are capable of N_2 assimilation. Fortunately, there exist unique groups of natural agents (microorganisms) for the conversion of this nitrogen to bioavailable forms, such as ammonia-N and nitrite/nitrate nitrogen, *etc.* with the help of specific biomolecules, mostly enzymes governed by remarkable genes and relevant components. The present chapter discusses the microbial species and their biomolecular aspects that are involved in the nitrogen cycle. The conversion of molecular nitrogen to biologically available nitrogen followed by the interconversion of its reactive forms requires specific organisms and environments at certain times for the transformation of nitrogen to other states.[1,2] Natural transformation of nitrogen includes six major stages, namely nitrogen fixation, nitrification, anammox, denitrification and ammonification.

(i) Nitrogen fixation

The conversion of molecular nitrogen to a biologically available chemical form of nitrogen, which is mediated by a few prokaryotes, is termed nitrogen fixation. The triple bond between nitrogen atoms provides high stability to the nitrogen molecule and therefore, there is a requirement of a significant amount of energy to break this bond. Scientists have reported that this energy is equivalent to the requirement of eight electrons and at least sixteen ATP molecules, which is responsible for limiting the number of nitrogen fixation-efficient organisms.[1]

Currently, due to the higher application need of nitrogen in mostly the agricultural field, artificial nitrogen conversion *via* the Haber–Bosch process and the utilization of this nitrogen has a huge market and there is inevitable discharge to the environment, which leads to environmental quality degradation and pollution. There is also a small amount of nitrogen fixation occurrence due to lightning and the combustion of fossil fuels.

$$N_2 + 8H_2 + 8e^- \rightarrow 2NH_3 + H_2$$

(ii) Nitrification

The two-step process of conversion of ammonia to nitrate *via* nitrite is termed as nitrification and is of high importance in the nitrogen cycle. Sergei Winogradsky, more than a century ago, first isolated chemolithoautotrophic bacteria that grew with energy received from ammonia or nitrite oxidation and identified two-step nitrification mediated by distinct groups of slow-growing organisms, ammonia-oxidizing bacteria (AOB) and nitrite-oxidizing bacteria (NOB).[3] Nitrification requires oxygen and hence aerobic conditions are a major influencing environmental factor for the prokaryotes and archaea carrying out the process. Different significant groups are involved in the two steps of nitrification. The first step is transformation of ammonia to nitrite aerobically through the formation of intermediate hydroxylamine by ammonia oxidizing bacteria from the genera *Nitrosomonas*, *Nitrosospira*, and *Nitrosococcus* and various ammonia-oxidizing archaea. This process occurs in the presence of two enzymes, namely ammonia monooxygenase and hydroxylamine oxidoreductase, and generates a small amount of energy.[4,5] These microbes are generally autotrophs, which use ammonia as a source of energy for carbon dioxide fixation during organic carbon synthesis, are termed as nitrosofiers, and are known to be relatively very slow growers.

The second step of nitrification is the oxidation of nitrite to nitrate by another significant group of prokaryotes, named as nitrite oxidizing bacteria, mostly from the genera of *Nitrospira*, *Nitrobacter*, *Nitrococcus*, and *Nitrospina*. The energy generation in this process is also very small and hence these are slow growth organisms. The oxidation of ammonia to nitrate completes the nitrification process.

$$NH_3 + O_2 \rightarrow NO_2^- + 3H^+ + 2e^- \quad \text{Ammonia oxidation}$$

$$NO_2^- + H_2O \rightarrow NO_3^- + 2H^+ + 2e^- \quad \text{Nitrite oxidation}$$

(iii) Anammox

Ammonia oxidation using nitrite as an electron acceptor to produce gaseous nitrogen as the end product by the anammox prokaryotes in anaerobic and anoxic conditions (in the absence of oxygen) is termed as the anammox process. Anammox prokaryotes mostly belong to the bacteria phyla Planctomycetes and the process was first discovered in wastewater treatment plants with anoxic bioreactors with bacterium *Brocadia anammoxidans*. The annamox prokaryotes are distributed in a wide range of habitats characterized by low-oxygen zones, including ocean, coastal and estuarine sediments, mangroves, and freshwater lakes. Anammox is duly responsible for a high amount of nitrogen loss, specifically in the ocean, and plays an important role in the nitrogen cycle.[6]

(iv) Denitrification

Denitrification is the conversion of nitrate (NO_3^-) to N_2 by denitrifying organisms capable of reducing NO_3^- *via* the formation of nitrite (NO_2^-), nitric oxide (NO) and nitrous oxide (N_2O) under oxygen-limiting conditions.[7]

$$NO_3^- \rightarrow NO_2^- \rightarrow NO \rightarrow N_2O \rightarrow N_2$$

The enzymes nitrate- (Nar/Nap), nitrite- (nirK/NirS), NO- (cNor/qNor), and nitrous oxide-reductase (NosZI and NosZII) facilitate denitrification. In the absence of the complete set of denitrifying enzymes, the denitrifiers often do not reach the end product N_2 and release intermediates such as nitrite into solution, as well as the greenhouse gases NO and N_2O to the atmosphere. Periplasmic or membrane-bound nitrate reductases NapA and NarG, respectively, execute dissimilatory reduction of NO_3^- to NO_2^-. The cd_1-containing (NirS in prokaryotes) or copper-containing (NirK in a few protists and fungi) nitrite reductases facilitate the second step of $NO_2^- \rightarrow NO$.[8] Nitric oxide reductase (Nor in prokaryotes), and an alternative enzyme (P450Nor in fungi) process additional reduction of NO to the greenhouse gas N_2O. Nitrous oxide reductase (NosZ) in prokaryotes furnishes the final step of N_2O to N_2 conversion.
Nitrate is exposed for metabolism (both assimilatory and dissimilatory) and stored in living cells. Ammonium is generated, followed by incorporation into biomass (for example, proteins and nucleic acids) in the case of assimilatory nitrate reduction. The dissimilatory nitrate reduction process involves the utilization of nitrate as an electron acceptor in the near absence of oxygen (oxic/anoxic conditions) or nitrate-free conditions (anaerobic conditions) and is known for providing microbes with environmental flexibility.[9]

(v) Ammonification
Nitrogen exists in the form of organic nitrogen in the excreta of organisms or in the dead biomass/tissues. Ammonification is the conversion of this organic nitrogen to ammonia by decomposers, including various fungi and prokaryotes, that make it available for producers (plants/microbes) for their growth.[9,10]

The atmospheric nitrogen therefore moves in its various forms in the biosphere before completing the cycle with its journey back, which is processed by different groups of microorganisms equipped with specific mechanisms making it possible to support the necessity of nitrogen for functioning as well as growth and development in an ecosystem.

However, the natural nitrogen cycle balance is often interrupted by the anthropogenic contribution of nitrogen (in agricultural fields as nutrients) that is discharged to ecosystems (aquatic ecosystems prone to eutrophication) or as oxidized products (to the atmosphere), which is responsible for pollution and can severely influence the environment and health.

7.1.1 Ecophysiology of Nitrogen Reacting Microbes

The complete cycle of nitrogen interlinks the atmosphere, lithosphere and biosphere and involves five major steps as discussed above, along with many sub-steps in it with multiple groups of microbes responsible for their individual activities. The biochemistry of the N cycle is a complex interplay among many microorganisms catalysing different reactions with four core reduction pathways (nitrogen fixation, assimilatory nitrate reduction, dissimilatory nitrate reduction, and ammonification) and two oxidation pathways (nitrification and anammox).

Table 7.1 presents active organisms and their responsible proteins for various functions in the N-cycle. Nitrogen fixation, the initial step of the N cycle, is predominantly carried out by prokaryotes (bacteria and archaea) equipped with nitrogenase enzyme complexes capable of reducing atmospheric N_2 to ammonia, useful elements for amino acids and other dynamic bio-compounds.[11,12]

The nitrification of ammonia is initiated by ammonia-oxidizing microorganisms (*e.g.*, *Nitrosomonas* and *Nitrosococcus*) and involves the oxidation of ammonia (NH_3) with oxygen into nitrite followed by the oxidation of nitrite into nitrate by nitrite-oxidizing microorganisms (*e.g.*, *Nitrobacter*, *Candidatus nitromaritima*, *Nitrolancea hollandica*, *Nitrotoga arctica*, *Nitrospira defluvii*, and *Nitrosomonas nitrosa* with cyanase-positive *N. moscoviensis*).[3,5,13]

Assimilatory and dissimilatory nitrate reduction for nitrate conversion to ammonia is carried out by prokaryotes as well as many widely distributed protists and eukaryotes as reported in current research.[7,9,14] A few organisms, well known for nitrate reduction and storage, are freshwater ciliate *Loxodes* sp., benthic diatom *Amphora coffeaeformis* and pelagic diatom *Thalassiosira weissflogii*, fungi *Aspergillus terreus*, *A. nidulans*, *Fusarium oxysporum* and *Cylindrocarpon tonkinense*, foraminifers (namely, *Allogromiida*, *Miliolida*, *Rotaliida* and *Textulariida*, *Bolivina plicata, Bolivina seminuda*, *Valvulineria cf. laevigata*, and *Stainforthia* sp.), and a few gromiids.[7]

Occasional nitrate storage is reported in chlorophyte *Dunaliella tertiolecta*, dinoflagellate *Amphidinium carterae*, haptophytes *Isochrysis galbana* and *Pavlova lutheri*, *etc.*[9] Notable denitrifiers are reported from the genera *Pseudomonas*, *Ralstonia*, *Alcaligenes*, *Paracoccus*, *Rhodobacter*, *Rubrivivax*, *Thauera*, *Burkholderia*, *Bacillus*, and *Streptomyces* available in various ecosystems, fungi *Fusarium oxysporum*, and foraminifera species (*e.g.*, *Buliminella tenuata*, *Virgulinella fragilis*, and *Globobulimina turgida*). The species *Shewanella baltica* and *Marinobacter* spp. are major contributors of denitrification in marine ecosystems.

Anammox is the oxidation of ammonium (NH_4^+) into dinitrogen gas (N_2) using nitrite as an electron acceptor and occurs in the anammoxosome-containing species. The anammoxosome is a membrane bound compartment inside the cytoplasm of anammox bacteria, such as Planctomycetes.[15] The decomposers are mainly associated with ammonification and include fungi and various protist groups.

7.2 Genomics of N-cycling Microbes

In total 47 gene subfamilies have been identified in the Pfam and uniport databases that target the N-cycle pathway, including nitrification, anammox, assimilatory nitrate reduction, dissimilatory nitrate reduction, denitrification

Table 7.1 Organisms and their working elements in the N-cycle.

N-cycle function	Organism	Responsible protein family
Nitrogen fixation	*Azotobacter vinelandii*	Nitrogenase molybdenum-iron protein alpha chain
	Azotobacter vinelandii	Nitrogenase molybdenum-iron protein beta chain
	Azotobacter vinelandii	Nitrogenase delta subunit
	Rhizobium leguminosarum	Nitrogenase iron protein Nif H
	Crocosphaera watsonii WH 8502	Nitrogenase-stabilizing/protective protein
Nitrification	*Nitrifier*	Ammonia monooxygenase subunit B (archaea)
		Ammonia monooxygenase subunit C (archaea)
		Nitrite oxidoreductase, alpha subunit
		Hydroxylamine dehydrogenase
		Ammonia monooxygenase subunit A (archaea)
		Nitrite oxidoreductase, beta subunit
Anammox	*Candidatus Brocadia fulgida*	Hydrazine synthase subunit B
	uncultured planctomycete	Hydrazine oxidoreductase
	Kuenenia stuttgartiensis	Hydrazine synthase subunit A
Assimilatory nitrate reduction	*Phormidium laminosum*	Ferredoxin-nitrite reductase
	uncultured bacterium	Assimilatory nitrate reductase
	Haloferax mediterranei	Cytochrome *b*-561
	Bacillus subtilis (*strain 168*)	Assimilatory nitrate reductase electron transfer subunit
	Azospirillum brasilense	Assimilatory nitrate reductase catalytic subunit
Dissimilatory nitrate reduction	*Escherichia coli (strain K12)*	Protein Nrf D
	Escherichia coli (*strain K12*)	Periplasmic nitrate reductase Nap A
	Bacillus subtilis (*strain 168*)	Nitrate reductase molybdenum cofactor assembly chaperone
	Escherichia coli	Nitrate reductase 2, beta subunit
	Candidatus Accumulibacter sp. *BA-91*	Nitrate reductase 2, alpha subunit
	Escherichia coli (*strain K12*)	Nitrite reductase (NADH) large subunit
		Nitrate reductase
		Nitrite reductase (NADH) small subunit
		Protein NrfC
		Nitrite reductase (cytochrome *c*-552)
		Cytochrome *c*-type protein NrfB
		Cytochrome *c*-type protein NapC
	Escherichia coli O157:H7	Nitrate reductase 2, delta subunit

Denitrification	*Escherichia coli* (*strain K12*)	Cytochrome *c*-type protein Nap C
	Cereibacter sphaeroides (*Q53177*)	Cytochrome *c*-type protein Nap B
	Cereibacter sphaeroides	Cytochrome *c*-type protein Nap B
	Paracoccus denitrificans (*strain Pd 1222*)	Nitrite reductase (NO-forming)
	Pseudomonas aeruginosa	Nitric oxide reductase subunit C
	Escherichia coli (*strain K12*)	Nitrate reductase 2, alpha subunit
	Pseudomonas stutzeri (*Pseudomonas perfectomarina*)	Nitrous-oxide reductase
	Pseudomonas stutzeri (*Pseudomonas perfectomarina*)	Nitric oxide reductase subunit B
	Escherichia coli O157:H7	Nitrate reductase 2, delta subunit
	—	Nitrate reductase
	*Klebsiella pneumoniae sub*sp. *pneumoniae*	Nitrate reductase 2, beta subunit
	Denitratisoma oestradiolicum	Nitrate reductase 2, gamma subunit
	Denitratisoma oestradiolicum	Nitrate reductase 2, gamma subunit
	Nitrosomonas europaea	Nitrite reductase (NO-forming)
	Escherichia coli (strain K12/DH10B)	Periplasmic nitrate reductase Nap A
	Escherichia coli (strain K12)	Nitrate reductase molybdenum cofactor assembly chaperone
		Nitrate reductase
		Nitrate reductase
		Nitrate reductase gamma subunit
		Nitrate reductase gamma subunit
Ammonification	*Klebsiella aerogenes* (*Enterobacter aerogenes*)	Urease subunit gamma
	Helicobacter pylori	Urease subunit beta
	Bacillus subtilis	Urease subunit alpha
	Pseudomonas aeruginosa	Nitronate monooxygenase
	Fusarium oxysporum	Nitroalkane oxidase

and ammonification performed by the microbes, and analyzed. The major aspects of the N-cycle and their responsible gene families are shown in Figure 7.1 as derived using Cytoscape software to generate the gene family and function network.

AnfG, nifD, nifH, nifK, and nifW are five gene families associated with nitrogen fixation (Noda *et al.*, 1999) and are responsible for regulating the synthesis and performance of the required necessary protein/enzyme. Similarly, for nitrification, gene families such as amoA, amoB, amoC, hao, nxrA, and nxrB are involved. A total of 16 gene families are recruited for denitrification, including narY, narV, nirK, napA, narJ, narH, narI, narZ, nosZ, norB, narG, napC, napB, nirS, norC and narW. All four phases of denitrification are controlled by these 16 gene families. For assimilatory nitrate reduction, six gene families, namely nasA, nasB, narB, narC, NR, and nirA, are generally well monitored. The genes nasAB, narBC, and NR are responsible for converting nitrate to nitrite, whereas nirA is responsible for converting nitrite to ammonia. For dissimilatory nitrate reduction, thirteen gene families including narZ, narY, nirB, narG, nirD, nrfC, nrfA, nrfB, napC, nrfD, napA, narJ, and narW, are responsible for nitrite reduction to ammonia as identified from the available literature. For anammox, three gene families: Hzo, kuste2861, and hdh have been identified. Among these, kuste2861 is involved in anaerobic ammonium oxidation, in which nitrite is employed as an electron acceptor in the conversion of ammonium to dinitrogen gas (N_2) and water, and hzo and hdh are involved in the oxidation of hydrazine to nitrogen.[16]

7.3 Nitrogen Reacting Microbes and Their Environmental Interactions

As mentioned above, nitrogen is exposed to aerobic oxidation through nitrification and anaerobic oxidation *via* ammonification and denitrification. The microbes efficient for nitrogen reaction are widely distributed and of diverse forms. Microorganisms play crucial roles in nitrogen transformations from the atmosphere to the biosphere as major operators of the N-cycle for supplementing the necessary N requirement.[17,18] From a similar perspective, these organisms are utilized for environmental purification and conservation. When discharged in high concentrations, nitrogen in its various forms is responsible for degrading the environment quality and is a major pollutant with various health impacts. Numerous microorganisms are also involved in the treatment of nitrogenous waste (in water/soil/air) according to their cellular and molecular composition. Nitrifiers equipped with the necessary enzymes are employed for the treatment of ammonia waste, and denitrifiers are exposed to nitrate/nitrite rich wastes. The nitrogen fixing microbes, for example, species from Rhizobium, are mostly found associated with leguminous plants that release root exudates sending signals to the microbes and the microbes have specific receptors for these signals so that they can respond and form symbiotic interactions.[19,20] The nitrifiers for

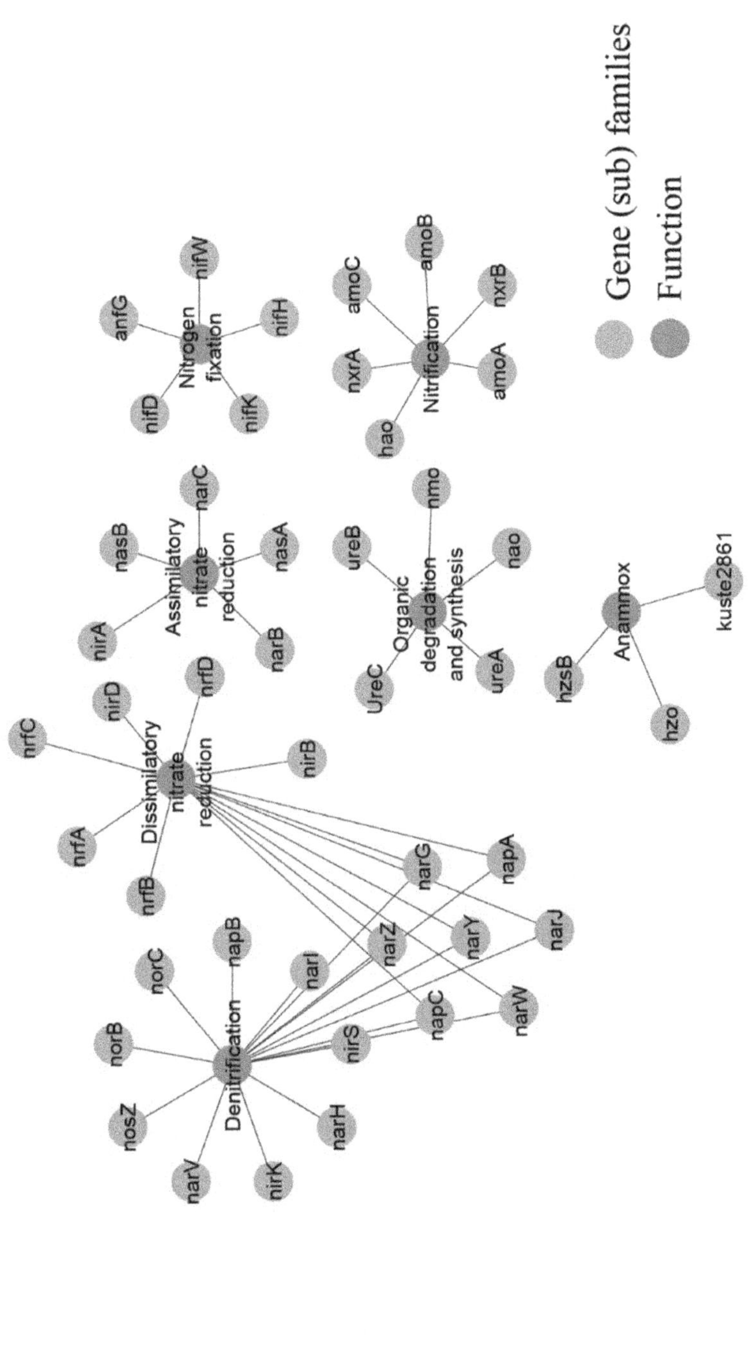

Figure 7.1 Aspects of the N-cycle and the involved significant gene families.

ammonia oxidization and nitrite oxidations are usually chemolithoautotrophs as well as photoautotrophs.[5] The former use other minerals whereas the latter use light as a source of energy for the oxidation of nitrogen. There occur natural interactions in terms of competition for necessary elements and habitat in the ecosystem. The nitrifiers are well known for their slow growth and therefore are always sensitive to the presence of heterotrophs present in the habitat. Water content and oxygen availability also play a key role in the activity and population of nitrifying microbes. The denitrifying organisms, on the other hand are more stable and robust in the presence of toxic pollutants compared to sensitive aerobic nitrifiers. Denitrifying organisms survive in microaerobic as well as anaerobic conditions and offer healthy competition with conventional heterotrophs. Many organisms involved in the N-cycle alter their responsivity according to the environment and chemical form of nitrogen they get exposed to. Many bacterial species have been discovered that are observed to perform alternate functions according to the environment that they are exposed to in different ecosystems. This is possible due to their specific protein/gene components that instruct the synthesis of the required molecules for a specific function.

7.4 Human Activity Swaying the Nitrogen Cycle

Aspects of all biogeochemical cycles are governed by mostly the sun and nature itself with ideal mechanisms and pathways since ancient times making ecosystem establishment, functioning and sustenance possible. Human activity branded as industrial and agricultural development has resulted in the introduction of many externally synthesized compounds widely responsible for impeding the natural balance of the ecosystem. From the atmosphere, the inaccessible nitrogen is converted through a limited number of nitrogen fixing organisms, making it available for usage by others. Nitrogen generated by the industrial nitrogen fixation method, the Haber–Bosch process, has increased exponentially since the 1940s, and anthropogenic activities are responsible for half, 210 Tg N of a total 413 Tg of reactive nitrogen, of global nitrogen fixation.[21–23] Compared to the global natural nitrogen fixation comprised by marine ecosystems, the human-contributed nitrogen fixation occurs on land, in the soil ecosystems, as it is highly used as fertilizers in agriculture for vegetation growth and land ecosystems are exposed to the majority of anthropogenic nitrogen transformation. The external addition of nitrogen is prone to disturb the nutrient balance in the soil and vegetation, thereby affecting ecosystem health.[24] Also, with increased availability of nitrogen, organisms are susceptible to more synthesis of organic compounds, hence influencing the carbon storage of the ecosystem. Over longer durations, there occurs an imbalance in the nutrients and a negative impact on biodiversity. From agricultural systems, unused nitrogen, mostly in the form of nitrate from fertilizers, gets released from the soil and makes its way with the runoff towards rivers or other aquatic ecosystems and often reaches drinking water. A higher concentration of nitrogenous nutrients results in eutrophication in the water body and is responsible for the death of aquatic ecosystems.[24,25] The presence of nitrate at a high level in drinking water is reported to cause blue baby disease in children. On the other hand, the release of nitrogen to the atmosphere in gaseous forms, NO, HNO_2, *etc.*, acts as a reactant for producing other secondary air pollutants (for example, PAN), enhancing phenomena such as acid rain, and climate change.[23,26] With the continuation of the human population explosion, human activities are continuously and severely threatening our resources and significantly altering the global nitrogen cycle, which could obliterate nature.

7.5 Conclusion

Nitrogen in its various forms possesses immense importance for the proper functioning of the biological world. All life is related to nitrogen as it is a major component of life elements, proteins, the molecular components and basic building blocks of organisms. Groups of organisms interact with nitrogen to transforms its form, while extracting energy for their survival and growth, thus providing the required nitrogen form to the next organisms. The various steps of the N-cycle, namely nitrogen fixation, nitrification, and denitrification, are microbially driven processes accountable for the bulk of nitrogen transformations, and various groups play their individual role in the fate of nitrogen in the Earth's ecosystems. External factors, such as the utilization of organic or chemical fertilization in the soil, the presence of toxic elements such as heavy metals, phenol compounds, and cyanates, *etc.*, negatively influence the nitrogen reacting microbes in waste soil/water treatment plants. The N reacting microbes are in the base line of the food chain of an ecosystem for nitrogen contribution of the biological nitrogen requirement, as well as in the end as decomposers in biological N recycling. Therefore, any negative affect on the nitrogen reacting organisms can cause total disturbance in the structure and function of an ecosystem.

References

1. B. Anne, *Nat. Educ. Knowl.*, 2010, **3**, 25.
2. L. Y. Stein and M. G. Klotz, *Curr. Biol.*, 2016, **26**, R94–R98.
3. H. Daims, S. Lücker and M. Wagner, *Trends Microbiol.*, 2016, **24**, 699.
4. J. Wang, X. Sun, Y. Xing, J. Xia and X. Feng, *J. Cleaner Prod.*, 2019, **240**, 118171.
5. M. Könneke, A. E. Bernhard, J. R. De La Torre, C. B. Walker, J. B. Waterbury and D. A. Stahl, *Nature*, 2005, **437**, 543–546.
6. M. M. M. Kuypers, G. Lavik, D. Woebken, M. Schmid, B. M. Fuchs, R. Amann, B. B. Jørgensen and M. S. M. Jetten, *Proc. Natl. Acad. Sci. U. S. A.*, 2005, **102**, 6478–6483.

7. C. Woehle, A. S. Roy, N. Glock, T. Wein, J. Weissenbach, P. Rosenstiel, C. Hiebenthal, J. Michels, J. Schönfeld and T. Dagan, *Curr. Biol.*, 2018, **28**, 2536.
8. S. W. Kim, S. Fushinobu, S. Zhou, T. Wakagi and H. Shoun, *Appl. Environ. Microbiol.*, 2009, **75**, 2652.
9. A. Kamp, S. Høgslund, N. Risgaard-Petersen and P. Stief, *Front. Microbiol.*, 2015, **6**, 1492.
10. M. Balk, A. M. Laverman, J. A. Keuskamp and H. J. Laanbroek, *Front. Microbiol.*, 2015, **6**, 166.
11. E. Kavehei, B. Shahrabi Farahani, G. A. Jenkins, C. Lemckert and M. F. Adame, *Water Res.*, 2021, **188**, 116511.
12. Y. E. Navarro-Noya, E. Hernández-Mendoza, J. Morales-Jiménez, J. Jan-Roblero, E. Martínez-Romero and C. Hernández-Rodríguez, *Appl. Soil Ecol.*, 2012, **62**, 52–60.
13. A. E. Bernhard, Z. C. Landry, A. Blevins, J. R. De La Torre, A. E. Giblin and D. A. Stahl, *Appl. Environ. Microbiol.*, 2010, **76**, 1285–1289.
14. T. E. Hanson, B. J. Campbell, K. M. Kalis, M. A. Campbell and M. G. Klotz, *Front. Microbiol.*, 2013, **4**, 180.
15. M. C. F. Van Teeseling, R. J. Mesman, E. Kuru, A. Espaillat, F. Cava, Y. V. Brun, M. S. Vannieuwenhze, B. Kartal and L. Van Niftrik, *Nat. Commun.*, 2015, **6**, 1–6.
16. B. Kartal, W. J. Maalcke, N. M. de Almeida, I. Cirpus, J. Gloerich, W. Geerts, H. J. M. Op den Camp, H. R. Harhangi, E. M. Janssen-Megens, K.-J. Francoijs, H. G. Stunnenberg, J. T. Keltjens, M. S. M. Jetten and M. Strous, *Nature*, 2011, **479**, 127–130.
17. K. Liao, H. Hu, S. Ma and H. Ren, *Water Res.*, 2019, **159**, 397–405.
18. N. Yang, C. Zhang, L. Wang, Y. Li, W. Zhang, L. Niu, H. Zhang and L. Wang, *Water Res.*, 2021, **206**, 117730.
19. K. K. Compton and B. E. Scharf, *Front. Plant Sci.*, 2021, **12**, 818.
20. B. Li, Y. Y. Li, H. M. Wu, F. F. Zhang, C. J. Li, X. X. Li, H. Lambers and L. Li, *Proc. Natl. Acad. Sci. U. S. A.*, 2016, **113**, 6496–6501.
21. P. M. Vitousek, D. N. L. Menge, S. C. Reed and C. C. Cleveland, *Philos. Trans. R. Soc., B*, 2013, **368**, 20130119.
22. D. Fowler, M. Coyle, U. Skiba, M. A. Sutton, J. N. Cape, S. Reis, L. J. Sheppard, A. Jenkins, B. Grizzetti, J. N. Galloway, P. Vitousek, A. Leach, A. F. Bouwman, K. Butterbach-Bahl, F. Dentener, D. Stevenson, M. Amann and M. Voss, *Philos. Trans. R. Soc., B*, 2013, **368**, 20130164.
23. W. San Martín, in *Life on Land*, ed. W. Leal Filho, A. M. Azul, L. Brandli, A. Lange Salvia and T. Wall, Springer International Publishing, Cham, 2020, pp. 1–16.
24. R. Xu, B. Li, E. Xiao, L. Y. Young, X. Sun, T. Kong, Y. Dong, Q. Wang, Z. Yang, L. Chen and W. Sun, *Environ. Pollut.*, 2020, **261**, 114226.
25. E. Broman, M. Zilius, A. Samuiloviene, I. Vybernaite-Lubiene, T. Politi, I. Klawonn, M. Voss, F. J. A. Nascimento and S. Bonaglia, *Water Res.*, 2021, **194**, 116954.
26. B. He, S. Kanae, T. Oki, Y. Hirabayashi, Y. Yamashiki and K. Takara, *Water Res.*, 2011, **45**, 2573–2586.

CHAPTER 8

Cell Biology, Biochemistry and Metabolism of Unique Anammox Bacteria

JYOTSNA CHOUBEY,[a] JYOTI KANT CHOUDHARI,[b] J. ANANDKUMAR,[c] MUKESH KUMAR VERMA,[c,d] TANUSHREE CHATERJEE[a] AND BIJU PRAVA SAHARIAH*[d]

[a] Raipur Institute of Technology Raipur, Chhattisgarh, 4911020, India; [b] Maulana Azad National Institute of Technology, Bhopal-462003, Madhya Pradesh, India; [c] National Institute of Technology Raipur, Chhattisgarh, 492010, India; [d] Chhattisgarh Swami Vivekanand Technical University, Bhilai, CG, India
*Email: biju.sahariah@gmail.com

8.1 Introduction

The anaerobic ammonium oxidation (anammox) process is an important microbial process in the nitrogen cycle that converts ammonia to N_2 gas by using nitrite as an electron acceptor under anaerobic conditions.[1,2] The anammox process takes place in many natural environments. Unlike conventional denitrification, the anammox process is cost effective and environmentally friendly as it does not require aeration and reduces the production of greenhouse gases (*i.e.*, N_2O and CO_2).[3,4] In the anammox reaction, nitrite and ammonium ions are converted directly into diatomic nitrogen and water. The bacteria that perform the annamox process were identified in 1999, and contribute ~23–30% of N loss from marine environments.[5–7] The contributions of anammox bacteria to biogeochemical N_2 production were measured as 18–36% in groundwater,[8] 9–13% in lakes[9]

Ammonia Oxidizing Bacteria: Applications in Industrial Wastewater Treatment
Edited by Maulin P. Shah

Published by the Royal Society of Chemistry, www.rsc.org

and 30–50% of molecular nitrogen formed in the oceans. These results show that anammox bacteria play a significant role in the global N cycle. Also, anammox bacteria show similarities to both Archaea and Eukarya, and are therefore interesting from an evolutionary perspective.

Anammox bacteria can grow under −2 to 43 °C temperatures and 6.7 to 8.3 pH levels. The optimal pH is 8, while the optimal temperature varies according to species.[10,11] Moreover, anammox bacteria are slow growing and require special enrichment and cultivation techniques.[12]

The bacteria that perform the anammox process belong to the bacterial phylum Planctomycetes, order "Candidatus Brocadiales", and family "Candidatus Brocadiaceae".[13] Up to now, five different genera have been described: "Candidatus Brocadia",[14] "Candidatus Kuenenia",[15] "Candidatus Scalindua",[16,17] "Candidatus Anammoxoglobus",[18] and "Candidatus Jettenia".[19]

So far, anammox species have not been isolated as classical pure cultures, lending them a Candidatus status. The genus *Scalindua* is mostly present in marine environments[18] or other environments with an enhanced salt concentration, and the other genera are mostly present in freshwater systems and soils. This chapter aims to focus on the physiology, cell biology, biochemistry and metabolism of unique anammox bacteria.

8.2 History of the Discovery of Anammox

Until the end of the twentieth century, it was widely considered that ammonium is an inert molecule in anoxic circumstances, and that activation by oxygen is essential for its metabolism, as it is for nitrifying bacteria. Broda (1977), an Austrian scientist, predicted the presence of microbes capable of ammonium oxidation with nitrite or nitrate as the electron acceptor based on thermodynamic calculations.[20] Richards (1965) reported an unexplained loss of ammonium under anoxic circumstances a decade earlier in studies of the nitrogen balance in anoxic fjords.[21]

A similar observation in a denitrifying bioreactor in Delft, The Netherlands,[22] sparked an extensive search for the bacteria responsible thirty years later. Strous *et al.* (1999) were the first to physically purify anammox cells from a laboratory enrichment culture, which led to the first description of an anammox bacteria. In the absence of oxygen, pure anammox cells turned ammonium and nitrite into dinitrogen gas and fixed cellular carbon from CO_2.[14] The first anammox bacterium was named "*Brocadia anammoxidans*", and it was awarded the rank of Candidatus because it was not pure by traditional microbiological standards.

Anammox bacteria have a complicated cell design with a central compartment, similar to that of other Planctomycetes (phylogenetically linked to anammox bacteria). Despite the fact that the detection of microbes capable of anaerobic ammonium growth was ineffective for a long time, numerous research groups are currently exploring various aspects of the anammox process.[23] Five anammox taxa have been identified so far, with 16S rRNA gene sequence identities ranging from 87 to 99 percent.[11] Despite their evolutionary distance, all anammox creatures are members of the same monophyletic cluster (order) known as the Brocadiales, which is connected to the Planctomycetales.

The Brocadiales are a branch of the Planctomycetes phylum.[14,24,25] "Kuenenia",[26,27] "Brocadia",[14,28,29] "Anammoxoglobus",[30] and "Jettenia"[29] are four "Candidatus"[19] and "Candidatus Scalindua"[16,17,31] is a fifth anammox genus that has been found in a variety of natural environments, including marine sediments and oxygen minimum zones (OMZs).[5,25,32,33]

8.3 The Natural Nitrogen Cycle and the Principle of the Anammox Process

Nitrogen is an essential component of all living beings as it is a key component of proteins and nucleic acids. It occurs in many forms and is continuously cycled in this form by a variety of bacteria. The nitrogen cycle was first discovered in 1857. It is the biogeochemical cycle in which nitrogen moves through both living and non-living things and is converted into various chemical compounds as it circulates among the atmosphere, and terrestrial and marine ecosystems. The nitrogen cycle includes important processes such as nitrogen fixation, ammonification, nitrification and denitrification (Figure 8.1).

After a few decades when the conventional nitrogen cycle was fully established, a new pathway in the nitrogen cycle was discovered by the

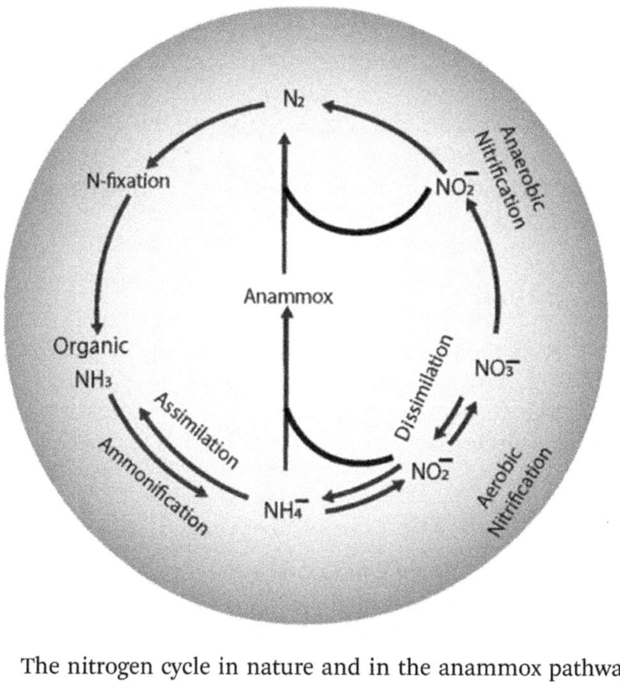

Figure 8.1 The nitrogen cycle in nature and in the anammox pathway.

research group of Arnold Mulder and Astrid van der Graaf from Delft University of Technology (Netherlands). In a denitrifying fluidized bed reactor, they observed that in anaerobic conditions, nitrogen gas was directly produced from ammonium (NH_4^+) and nitrite (NO_2^-) by a new species of denitrifying bacteria. This bypasses the intermediate step of nitrate formation.[22] These newly identified denitrifying bacteria became pioneers for research studies into their characteristics and application. Later, it became a milestone in various industrial applications, especially in waste water treatment. The study was published in the Nature Journal with a commercial name Anammox (ANaerobic AMMonium OXidation).[14] The Anammox pathway, which is an anaerobic ammonium oxidation process, involves two important biological processes: the catabolic (dissimilation) and anabolic (assimilation) reactions.

1. The process of the catabolic reaction (dissimilation):

$$NH_4^+ + NO_2^- \rightarrow N_2 + 2H_2O$$

2. The process of the anabolic reaction (assimilation):

$$HCO_3^- + 0.2NH_4^+ + 2.1NO_2^- + 0.8H^+$$

$$\rightarrow CH_{1.08}O_{0.5}N_{0.2} + 2.1NO_3^- + 0.4H_2O$$

8.4 Cell Biology of Anammox Bacteria

Anammox bacteria are coccoid with a diameter less than 1.0 mm. The generation time is 10–30 days and they are physiologically distinct from the other known Planctomycetes: they are anaerobic chemolithoautotrophs. However, under an electron microscope, they show a quite complicated structure that is a unique characteristic of the Planctomycetes discussed. All anammox bacteria contain three membrane-bound compartments in their cell (Figure 8.2) which is discussed below.

8.4.1 The Cell Wall

They possess a proteinaceous cell wall devoid of both peptidoglycan and outer membrane. So, they are neither Gram-positive nor Gram-negative, as is proposed for all Planctomycetes.[34] There are some indications of the presence of an S-layer protein lattice in *K. stuttgartiensis*.[35]

8.4.2 The Paryphoplasm

The outer compartment of the anammox cell is called the paryphoplasm, which accounts for on average 20% of the total cell volume and has an unknown function. It is bound by the cytoplasmic membrane, which defines the cell boundary. Ladderane lipids present in the cytoplasmic membrane

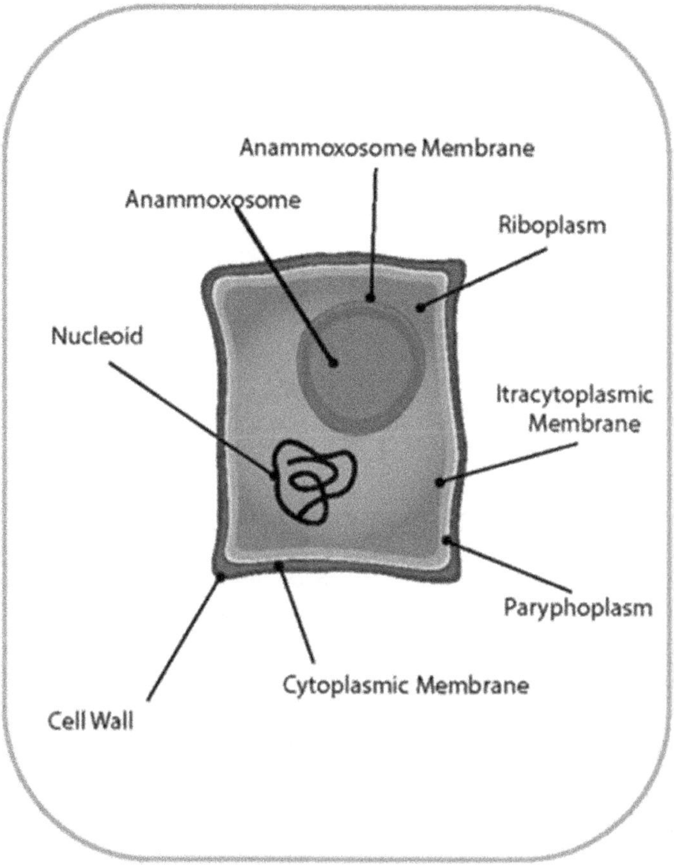

Figure 8.2 Cell plan of the anammox cell showing the three different compartments and their surrounding membranes.

may provide additional strength. A special structure, a non-FtsZ-based cell division ring, is found in the paryphoplasm, which is striking as the anammoxosome also divides during cell division and no membrane links have been observed between the anammoxosome and the other anammox membranes.[36]

8.4.3 The Riboplasm

The riboplasm is the middle compartment of the anammox cell and accounts for an average of 20% of the total cell volume. It is surrounded by the intracytoplasmic membrane and contains the nucleoid, ribosomes and vesicles. Therefore, it is the site of transcription, translation and glycogen storage.[37] In anammox bacteria, the building blocks for protein synthesis are derived *via* autotrophic carbon fixation, which occurs through the acetyl-CoA (Wood–Ljungdahl) pathway as suggested from the metagenome of *K. stuttgartiensis*. Proteins synthesized in the riboplasm have to be transported to another compartment of the cell. For this, both the Sec (secretory) pathway and TAT (twin arginine translocation) system are involved in the transport of proteins in annamox bacteria.[27] Furthermore, it was suggested that the TAT system might be specifically located in the anammoxosome membrane and the Sec pathway would assist in protein transport towards all compartments. Yet transportation of proteins from the riboplasm to either the paryphoplasm or anammoxosome remains unclear.[38]

8.4.4 The Anammoxosome

The anammoxosome is the innermost and largest compartment of the anammox cell. It accounts for an average of 60% of the total cell volume. It is inherited to the daughter cell during cell division, highly suggesting it to be a true cell organelle. All catabolic processes of anammox metabolism occur in the anammoxosome. The anammoxosome membrane is a highly curved membrane that surrounds the anammoxosome and is thought to house several enzymes involved in the electron transport chain, proton motive force generation, and subsequent ATP synthesis.[39,40]

The anammoxosome harbours the energy metabolism of the anammox cell and is therefore similar in function to the Eukaryotic mitochondria. It is the first and only energy-conserving cell compartment in prokaryotes that has been discovered to date. The anammoxosome membrane consists of specialized ladderane lipids,[41] which are only found in the annamox group of microorganisms, forming a proton gradient between the anammoxosome and paryphoplasm.

8.5 Biochemistry of the Anammox Reaction

The anammox process is very interesting in its biochemical aspects, because of the way that the inert ammonium molecule is handled and the way ATP is conserved. In the anammox reaction, nitrite serves as the terminal electron acceptor and produces dinitrogen gas. This process has only been discovered in anammox bacteria so far. The anammox process comprises three redox reactions: (i) nitrite (NO_2^-) is reduced to nitric oxide (NO), (ii) nitric oxide is coupled with ammonium (NH_4) to produce hydrazine (N_2H_4) and (iii) dinitrogen (N_2) gas is produced when hydrazine (N_2H_4) is oxidised. The enzymes that catalyse nitrite reduction are still a point of contention. It has been observed that in different anammox species, different enzymes are responsible for the anammox reaction.[42]

(1) $NO_2^- + 2H^+ + e^- \rightarrow NO + H_2O$
(2) $NO + NH_4^+ + 2H^+ + 3e^- \rightarrow N_2H_4 + H_2O$
(3) $N_2H_4 \rightarrow N_2 + 4H^+ + 4e^-$

Hydrazine, a highly energetic intermediate produced only by anammox bacteria, is produced by the enzyme hydrazine synthase (HZS). Hydrazine dehydrogenase catalyses the oxidation of hydrazine (HDH) to dinitrogen. The electrons created in the reaction are thought to create a proton motive force (pmf) that is used to produce ATP by membrane bound ATP synthase.

Furthermore, anammox bacteria oxidize large amounts of nitrite to nitrate (NO_3^-) and it has been suggested that this reaction produces electrons that act as reducing agents for carbon dioxide (CO_2) fixation *via* the acetyl-CoA pathway.[10,27,43,42] However, carbon fixation involves several low redox potential reactions, and the redox potential of electrons donated by nitrite oxidation is significantly higher. It was recently shown that anammox bacteria can also grow without oxidizing nitrite when cultured in the presence of nitric oxide. This suggests that electrons generated by nitrite oxidation are not used for carbon fixation, but for nitrite reduction to nitric oxide.[42,44]

8.6 Application of the Anammox Process

In recent years, research efforts have been dedicated to the more applied aspects of the anammox process. The most important applications of the Anammox process are for wastewater treatment; for example, from the landfill leachate, monosodium glutamate wastewater, and pharmaceutical wastewater. These types of wastewater contain an extremely high content of NH_4^+. In contrast to the domestic wastewater, it has only a small amount of nitrogenous substances. The anammox process of nitrogen removal reduces operational costs by 90%. The anammox process replaces the conventional denitrification step in wastewater treatment, if preceded by partial nitrification to nitrite, and thereby reduces nitrification aeration costs by 50%. In the first step of partial nitrification (nitritation), half of the ammonium is converted to nitrite by ammonia oxidizing bacteria:

$$2NH_4^+ + 3O_2 \rightarrow 2NO_2^- + 4H^+ + 2H_2O$$

The resulting ammonium and nitrite are converted in the anammox process to dinitrogen gas and ~15% nitrate (not shown) by anammox bacteria:

$$NH_4^+ + NO_2^- \rightarrow N_2 + 2H_2O$$

Both processes can take place in 1 reactor, where two guilds of bacteria form compact granules.

8.7 The Anammox Metabolism

Anammox bacteria use CO_2/bicarbonate as their only carbon source to convert ammonium and nitrite. The substances are digested in lab-scale bioreactors under steady-state conditions according to this overall eqn (8.1):[45]

$$1NH_4^+ + 1.32NO_2^- + 0.066HCO_3^- + 0.13H^+ \rightarrow 1.02N_2 + 0.26NO_3^- + 0.066CH_2O_{0.5}N_{0.15} + 2.03H_2O \quad (8.1)$$

$$NH_4^+ + 1NO_2^- \rightarrow 1N_2 + 2H_2O\left(DG°0 = -357\ \text{kJ mol}^{-1}\ N_2\right) \quad (8.2)$$

$$0.26NO_2^- + 0.066HCO_3^- \rightarrow 0.26NO_3^- + 0.066CH_2O_{0.5}N_{0.15} \quad (8.3)$$

The overall reaction is the sum of two partial reactions: the energy-generating process (eqn (8.2)), which involves ammonium oxidation coupled to nitrite reduction to produce dinitrogen gas, and bicarbonate fixation into cell biomass ($CH_2O_{0.5}N_{0.15}$) (eqn (8.3)). Both nitrogen atoms, ammonium and nitrite, are derived from two separate substrates. This is useful for physiological and biochemical investigations, as well as for detecting anammox activity in natural systems. The particular end product of the anammox reaction when using ^{15}N-labeled substrates ($^{15}NH_4$ or $^{15}NO_2$) is $^{29}N_2$. Mass spectroscopy can be used to determine the isotope makeup of the produced dinitrogen gas ($^{14}N^{14}N$, $^{14}N^{15}N$, $^{15}N^{15}N$).[46]

Next, nitrite has a dual purpose: it serves as an electron acceptor in the ammonium oxidation reaction (eqn (8.2)) and as an electron donor in the CO_2 reduction to biomass reaction (eqn (8.3)). In the latter situation, nitrite is anaerobically converted to nitrate, and as a result, nitrate production is always linked to growth. 1 mol of carbon is bound per 15 catabolic cycles, according to the aforementioned chemical stoichiometries (eqn (8.1)–(8.3)). Similarly, per mol of fixed carbon, around 4 mol of nitrite is oxidised.

8.8 Conclusions

This chapter highlights the understanding of the environmental importance of the anammox bacteria in the global nitrogen cycle. Anammox bacteria are very interesting microbes that challenge many biological concepts. They play a significant role in the loss of nitrogen from the ocean and are useful for wastewater treatment. They have unique metabolisms with unusual intermediates, such as rocket fuel hydrazine, and carry out this reaction and subsequent energy saving in the so-called prokaryotic organelle, the anammoxosome. Anammox cells are separated by three separate lipid bilayer membranes containing unique ladder-like membrane lipids. Anammox bacteria offer an economically attractive and environmentally friendly alternative to current wastewater treatment processes for the removal of fixed nitrogen. The anammox process does not necessitate the input of carbon nutrients from outside the system. This method also creates less sludge and consumes less energy and oxygen.

Finally, anammox bacteria have proven to be of considerable (micro) biological interest throughout the last decade.

References

1. K. R. Arrigo, *Nature*, 2005, **437**(7057), 349–355.
2. A. A. van de Graaf, A. Mulder, P. de Bruijn, M. S. Jetten, L. A. Robertson and J. G. Kuenen, *Appl. Environ. Microbiol.*, 1995, **61**, 1246–1251.
3. M. S. M. Jetten, S. Logemann, G. Muyzer, L. A. Robertson, S. de Vries, M. C. M. van Loosdrecht and J. G. Kuenen, *Antonie van Leeuwenhoek*, 1997, **71**, 75–93.
4. U. van Dongen, M. S. M. Jetten and M. C. M. van Loosdrecht, *Water Sci. Technol.*, 2001, **44**, 153–160.
5. T. Dalsgaard, B. Thamdrup and D. E. Canfield, *Res. Microbiol.*, 2005, **156**, 457–464.
6. A. R. Babbin, R. G. Keil, A. H. Devol and B. B. Ward, *Science*, 2014, **344**, 406–408.
7. M. Trimmer and P. Engström, *Nitrification*, Wiley Online Library, 2011, pp. 201–235.
8. T. A. Moore, Y. Xing, B. Lazenby, M. D. J. Lynch, S. Schiff, W. D. Robertson, R. Timlin, S. Lanza, M. C. Ryan and R. Aravena, *Environ. Sci. Technol.*, 2011, **45**, 7217–7225.
9. C. J. Schubert, E. Durisch-Kaiser, B. Wehrli, B. Thamdrup, P. Lam and M. M. M. Kuypers, *Environ. Microbiol.*, 2006, **8**, 1857–1863.
10. B. Kartal, W. Geerts, M. S. M. Jetten, in *Methods in Enzymology*, ed. M. G. Klotz, Academic Press, 2011, vol. 486, pp. 89–108.
11. M. S. M. Jetten, L. van Niftrik, M. Strous, B. Kartal, J. T. Keltjens and H. J. M. Op den Camp, *Crit. Rev. Biochem. Mol. Biol.*, 2009, **44**, 65–84.
12. L. van Niftrik and M. S. M. Jetten, *Microbiol. Mol. Biol. Rev.*, 2012, **76**, 585–596.
13. M. S. M. Jetten, H. J. M. O. den Camp, J. G. Kuenen, M. Strous, *Bergey's Manual of Systematics of Archaea and Bacteria*, 2015, pp. 1–10.
14. M. Strous, J. A. Fuerst, E. H. M. Kramer, S. Logemann, G. Muyzer, K. T. van de Pas-Schoonen, R. Webb, J. G. Kuenen and M. S. M. Jetten, *Nature*, 1999, **400**, 446–449.
15. M. Schmid, U. Twachtmann, M. Klein, M. Strous, S. Juretschko, M. Jetten, J. W. Metzger, K.-H. Schleifer and M. Wagner, *Syst. Appl. Microbiol.*, 2000, **23**, 93–106.
16. M. Schmid, K. Walsh, R. Webb, W. I. Rijpstra, K. van de Pas-Schoonen, M. J. Verbruggen, T. Hill, B. Moffett, J. Fuerst and S. Schouten, *Syst. Appl. Microbiol.*, 2003, **26**, 529–538.
17. M. M. M. Kuypers, A. O. Sliekers, G. Lavik, M. Schmid, B. B. Jørgensen, J. G. Kuenen, J. S. Sinninghe Damsté, M. Strous and M. S. M. Jetten, *Nature*, 2003, **422**, 608–611.

18. B. Kartal, J. Rattray, L. A. van Niftrik, J. van de Vossenberg, M. C. Schmid, R. I. Webb, S. Schouten, J. A. Fuerst, J. S. Damsté and M. S. M. Jetten, *Syst. Appl. Microbiol.*, 2007, **30**, 39–49.
19. Z.-X. Quan, S.-K. Rhee, J.-E. Zuo, Y. Yang, J.-W. Bae, J. R. Park, S.-T. Lee and Y.-H. Park, *Environ. Microbiol.*, 2008, **10**, 3130–3139.
20. E. Broda, *Z. Allg. Mikrobiol.*, 1977, **17**, 491–493.
21. F. A. Richards, *Chem. Oceanogr.*, 1965, **1**, 611–643.
22. A. Mulder, A. A. van de Graaf, L. A. Robertson and J. G. Kuenen, *FEMS Microbiol. Ecol.*, 1995, **16**, 177–183.
23. J. G. Kuenen, *Nat. Rev. Microbiol.*, 2008, **6**, 320–326.
24. M. C. Schmid, B. Maas, A. Dapena, K. van de Pas-Schoonen, J. van de Vossenberg, B. Kartal, L. van Niftrik, I. Schmidt, I. Cirpus, J. G. Kuenen, M. Wagner, J. S. Sinninghe Damsté, M. Kuypers, N. P. Revsbech, R. Mendez, M. S. M. Jetten and M. Strous, *Appl. Environ. Microbiol.*, 2005, **71**, 1677–1684.
25. M. C. Schmid, N. Risgaard-Petersen, J. van de Vossenberg, M. M. M. Kuypers, G. Lavik, J. Petersen, S. Hulth, B. Thamdrup, D. Canfield, T. Dalsgaard, S. Rysgaard, M. K. Sejr, M. Strous, H. J. M. Op Den Camp and M. S. M. Jetten, *Environ. Microbiol.*, 2007, **9**, 1476–1484.
26. M. Schmid, U. Twachtmann, M. Klein, M. Strous, S. Juretschko, M. Jetten, J. W. Metzger, K.-H. Schleifer and M. Wagner, *Syst. Appl. Microbiol.*, 2000, **23**, 93–106.
27. M. Strous, E. Pelletier, S. Mangenot, T. Rattei, A. Lehner, M. W. Taylor, M. Horn, H. Daims, D. Bartol-Mavel, P. Wincker, V. Barbe, N. Fonknechten, D. Vallenet, B. Segurens, C. Schenowitz-Truong, C. Médigue, A. Collingro, B. Snel, B. E. Dutilh, H. J. M. Op den Camp, C. van der Drift, I. Cirpus, K. T. van de Pas-Schoonen, H. R. Harhangi, L. van Niftrik, M. Schmid, J. Keltjens, J. van de Vossenberg, B. Kartal, H. Meier, D. Frishman, M. A. Huynen, H.-W. Mewes, J. Weissenbach, M. S. M. Jetten, M. Wagner and D. le Paslier, *Nature*, 2006, **440**, 790–794.
28. J. G. Kuenen, *ASM News*, 2001, **67**, 456–463.
29. B. Kartal, L. van Niftrik, J. Rattray, J. L. C. M. van de Vossenberg, M. C. Schmid, J. Sinninghe Damsté, M. S. M. Jetten and M. Strous, *FEMS Microbiol. Ecol.*, 2008, **63**, 46–55.
30. B. Kartal, J. Rattray, L. A. van Niftrik, J. van de Vossenberg, M. C. Schmid, R. I. Webb, S. Schouten, J. A. Fuerst, J. S. Damsté, M. S. M. Jetten and M. Strous, *Syst. Appl. Microbiol.*, 2007, **30**, 39–49.
31. J. van de Vossenberg, J. E. Rattray, W. Geerts, B. Kartal, L. van Niftrik, E. G. van Donselaar, J. S. Sinninghe Damsté, M. Strous and M. S. M. Jetten, *Environ. Microbiol.*, 2008, **10**, 3120–3129.
32. C. R. Penton, A. H. Devol and J. M. Tiedje, *Appl. Environ. Microbiol.*, 2006, **72**, 6829–6832.
33. D. Woebken, P. Lam, M. M. M. Kuypers, S. W. A. Naqvi, B. Kartal, M. Strous, M. S. M. Jetten, B. M. Fuchs and R. Amann, *Environ. Microbiol.*, 2008, **10**, 3106–3119.

34. E. König, H. Schlesner and P. Hirsch, *Arch. Microbiol.*, 1984, **138**, 200–205.
35. J. A. Fuerst, R. I. Webb, L. van Niftrik, M. S. M. Jetten, M. Strous, in *Complex Intracellular Structures in Prokaryotes*, ed. J. M. Shively, Springer, Berlin, Heidelberg, 2006, pp. 259–283.
36. L. van Niftrik, W. J. C. Geerts, E. G. van Donselaar, B. M. Humbel, R. I. Webb, H. R. Harhangi, H. J. M. Op Den Camp, J. A. Fuerst, A. J. Verkleij, M. S. M. Jetten and M. Strous, *Mol. Microbiol.*, 2009, **73**, 1009–1019.
37. M. R. Lindsay, R. I. Webb, M. Strous, M. S. Jetten, M. K. Butler, R. J. Forde and J. A. Fuerst, *Arch. Microbiol.*, 2001, **175**, 413–429.
38. M. H. Medema, M. Zhou, S. A. F. T. van Hijum, J. Gloerich, H. J. C. T. Wessels, R. J. Siezen and M. Strous, *BMC Genomics*, 2010, **11**, 299.
39. L. van Niftrik, M. van Helden, S. Kirchen, E. G. van Donselaar, H. R. Harhangi, R. I. Webb, J. A. Fuerst, H. J. M. Op den Camp, M. S. M. Jetten and M. Strous, *Mol. Microbiol.*, 2010, **77**, 701–715.
40. N. de Almeida, S. Neumann and R. J. Mesman, *J. Bacteriol.*, 2015, **197**(14), 2432–2441.
41. H. A. Boumann, P. Stroeve, M. L. Longo, B. Poolman, J. M. Kuiper, E. C. Hopmans, M. S. M. Jetten, J. S. Sinninghe Damsté and S. Schouten, *Biochim. Biophys. Acta, Biomembr.*, 2009, **1788**, 1452–1457.
42. B. Kartal, N. M. de Almeida, W. J. Maalcke, H. J. M. Op den Camp, M. S. M. Jetten and J. T. Keltjens, *FEMS Microbiol. Rev.*, 2013, **37**, 428–461.
43. S. Schouten, M. Strous, M. M. M. Kuypers, W. I. C. Rijpstra, M. Baas, C. J. Schubert, M. S. M. Jetten and J. S. Sinninghe Damsté, *Appl. Environ. Microbiol.*, 2004, **70**, 3785–3788.
44. Z. Hu, PhD thesis, Radboud University Nijmegen, 2014.
45. M. Strous, J. J. Heijnen, J. G. Kuenen and M. S. M. Jetten, *Appl. Microbiol. Biotechnol.*, 1998, **50**, 589–596.
46. A. A. van de Graaf, P. de Bruijn, L. A. Robertson, M. S. M. Jetten and J. G. Kuenen, *Microbiology*, 1997, **143**, 2415–2421.

CHAPTER 9

Microbial Nitrogen Transformation and Recovery in Wastewater: Current Strategies and Applications

RITESH PATTNAIK,*[a] LOPAMUDRA RAY,[a,b] KOUSTAV SAHA,[a] SANJANA PAL[a] AND HARSHITA SINGH[a]

[a] School of Biotechnology, Kalinga Institute of Industrial Technology (Deemed-to-be-University), Bhubaneswar, Odisha, India; [b] School of Law, Kalinga Institute of Industrial Technology (Deemed-to-be-University), Bhubaneswar, Odisha, India
*Email: riteshpattnaik@gmail.com

9.1 Introduction

The continuous input of nitrogen (N) to water bodies like ponds, lakes and other aquatic systems is a major responsible factor for environment pollution and an emerging cause of eutrophication. Nitrogen removal from water bodies, including wastewater reservoirs, has become a serious global issue. The excessive occurrence of untreated reactive nitrogen can result in environmental degradation. The reactive nitrogen in the atmosphere not only promotes smog and fog generation, but also contributes towards ozone depletion. Acid deposition of reactive nitrogen from the atmosphere causes acidification of soils, and inadvertent fertilization of trees and grasslands. This hampers the nutrient balance in the soil, and hence decreases the ecosystem health.[1] Furthermore, issues like nitrate leaching may cause groundwater pollution with nitrate, posing serious health hazards, including

Ammonia Oxidizing Bacteria: Applications in Industrial Wastewater Treatment
Edited by Maulin P. Shah

Published by the Royal Society of Chemistry, www.rsc.org

the risk of Hodgkin lymphoma and colorectal cancer, when ingested by human beings.[2]

The conventional wastewater treatment plants (WWTPs) achieve nitrogen removal *via* physico-chemical or biological approaches. Though chemical processing requires a smaller reactor volume and facilitates mineralization of non-biodegradable compounds, its disadvantages often outweigh its benefits, fostering an increasing shift to more sustainable and environment-friendly practices of biological wastewater treatment.[3] Currently, pre-denitrification in activated sludge is considered as an ideal and common technique employed for nitrogen removal from wastewater units. Extensive research attention is focused on enhancing the efficiency of nitrogen removal through biological treatment techniques involving the fundamentals of the nitrogen cycle. The two main strategies for the removal of nitrogen from wastewater include conversion of inorganic nitrogen present in the wastewater into dinitrogen gas, or its concentration and recovery as fertilizer.[4] Nitrogen fertilization production accounts for 1–2% of global energy consumption.[5] Evading this energy consumption through direct conversion of wastewater nitrogen to fertilizers is a tempting, yet practically inefficient strategy due to the diluted concentrations of nitrogen in wastewater and high energy demand of the process.

The conventional process of nitrogen removal from wastewater involves nitrification–denitrification. Ammonia present in the wastewater is oxidized to nitrite and then to nitrate under aerobic conditions. Later, the oxidized nitrate is anaerobically converted to nitrogen gas in the presence of organic carbons as a source of electrons. However, it is economically expensive due to the high oxygen demand for complete nitrification, the requirement of an external carbon source for heterotrophic denitrifying microbes, high sludge production and greenhouse gas (N_2O) emission. To combat this economical hurdle, anammox-based technologies are employed that minimize the process by directly converting ammonia and nitrites into nitrogen gas.[6,7] Besides microbial techniques, processes like phytoremediation have also gained interest as an emerging green technology approach in wastewater treatment due to minimal operating costs and eco-friendliness.[8]

9.2 Biological Nitrogen Cycle

Nitrogen is an indispensable element for all living forms, and an integral part of key cell components such as proteins, RNA and DNA. Though it accounts for 78% of the atmosphere, it cannot be utilized by organisms directly due to the triple bond between the two nitrogen atoms, and is therefore limited for living organisms. Therefore, this non-reactive form of nitrogen needs to be converted to its reactive form (NH_3 or NH_4) after being fixed by specialized microorganisms or by industrial processes.[9] Most of the fixed nitrogen is generated biologically from a limited group of prokaryotes called diazotrophs that exhibit a unique enzymatic pathway converting N_2 into NH_3, among which some of these prokaryotic organisms show symbiosis relationships with legumes and plants, while others live freely in aquatic and terrestrial environments. There are quite a large number of nitrogenous compounds existing in nature with various oxidation states ranging from −3 to +5. The global nitrogen cycle (Figure 9.1) is considered to be the process where atmospheric N is brought into the terrestrial and marine ecosystems through biological nitrogen fixation (BNF), in which unreactive molecular nitrogen is reduced to ammonium compounds, and eventually the fixed nitrogen returns back into molecular nitrogen. Around 120 megatons of bioavailable nitrogen are delivered to the biosphere by means of natural processes, mostly due to nitrogen fixation by microbes and some in association with leguminous plants, namely alfalfa, clover, beans, peas, lupins and lentils. Currently, the industrially designed Haber–Bosch process has put these natural processes in the backseat, where N_2 and H_2 get converted to ammonia, thereby contributing roughly the same amount of fixed nitrogen as done by natural processes, around 100 million tons of nitrogen fertilizer each year. Recent estimates show that 40% of the human population depends on the contribution to the nitrogen cycle by humans. There exist two large nitrogen pools on the Earth – the biologically reactive nitrogen pool and the atmospheric nitrogen pool, among which internal interactions occur within the biologically reactive nitrogen pool, whereas in and out flow occur between the atmospheric and the reactive nitrogen pools. Only two major microbial processes, nitrogen fixation and denitrification, control the interrelationship between the two large nitrogen pools.

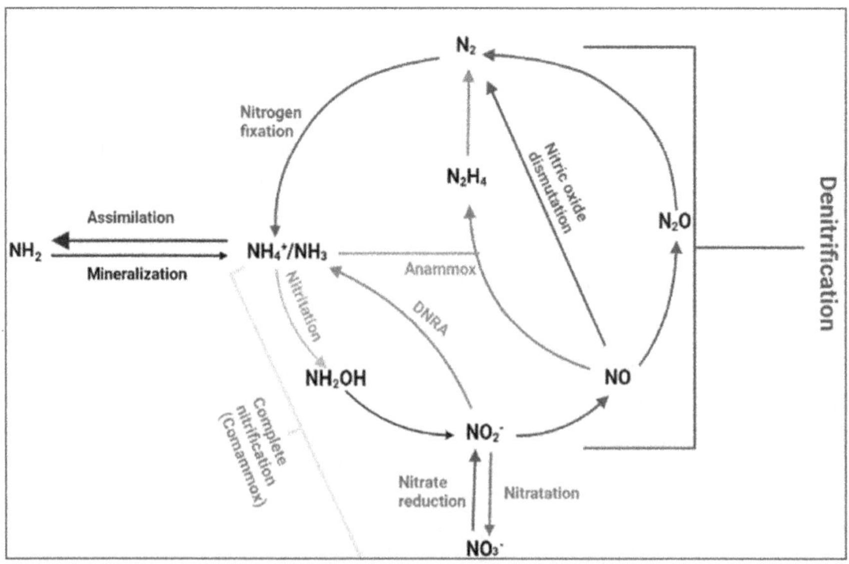

Figure 9.1 The typical biological nitrogen cycle showing all the reactions comprising the seven different processes.

In nutrient cycling of nitrogen, the unreactive dinitrogen species are fixed by nitrogen-fixing bacteria (NFB) into a much more usable form, like ammonia (NH_3). Ammonia is easily taken up by plants, the primary producers. Plants become the source of nitrogen for other organisms like animals. Ammonia is furthermore converted into other inorganic nitrogen compounds *e.g.* nitrate (NO_3^-) or nitrite (NO_2^-), as suited by the organism. Nitrate is then converted back to diatomic nitrogen gas (N_2), hence completing the cycle. These transformations are carried out by a variety of microorganisms, including bacteria, archaea, and fungi.

9.3 Imbalances of Excessive Nitrogen

A balanced nitrogen cycle is mainly dependent on collective transformation activities, like nitrogen fixation, nitrification, denitrification, nitrogen assimilation and dissimilation. However, increased urbanization and industrialization have led to distortion in the natural balance of the nitrogen cycle, causing accumulation of nitrogenous compounds in different ecosystems, including rivers and lakes. Excessive nitrogen and phosphorous also tend to impact public health and cause eutrophication *via* the excessive loss of available oxygen and causing other undesirable changes in the aquatic ecological system. To safeguard environment health, the exploitation of biological mechanisms seems essential within the nitrogen cycle in order to eliminate any accumulated nitrogen species, like ammonia and nitrate, and thereby assist in restoring the ecological balance of the nitrogen cycle. The efficient removal of harmful nitrogenous compounds to harmless nitrogen is subsequently discussed describing the mechanisms of nitrification and denitrification and advanced innovative technology-based approaches such as the anammox method, completely autotrophic nitrogen removal over nitrite (CANON) and oxygen-limited autotrophic nitrification and denitrification (OLAND) processes for nitrogen removal from wastewater.

9.4 Associated Microorganisms for Effective Nitrogen Removal

The biological nitrogen removal process occurs mainly with the release of N_2O and primarily follows the same nitrogen transformation processes of undergoing the transition of aerobic or anoxic reactions carried out by the indigenous microbial cultures existing in other environments, such as the soil, water, freshwater habitats and marine habitats. The two-step process of nitrogen removal from wastewater occurs initially with the nitrification process under strict aerobic conditions, followed by the denitrification process under anoxic conditions. Obligate aerobic autotrophs, known as ammonia oxidizing bacteria (AOB), such as *Nitrosomonas*. *Nitrosococcus* and *Nitrosospira* belonging to the β-proteobacteria lineage, oxidize the ammonia primarily present in wastewater to nitrite and this is eventually converted to nitrate by nitrite-oxidizing bacteria (NOB) such as *Nitrobacter*, *Nitrospina* and *Nitrocystis* from the α-proteobacteria lineage. The generation of N_2O is caused by the activities of AOB and NOB under aerobic conditions in wastewaters through nitrification–denitrification and incomplete oxidation of hydroxylamine (NH_2OH).[10] The biological denitrification process in wastewater under oxic/anoxic conditions is usually mediated by facultative anaerobic representatives of α, β and γ proteobacteria, like *Paracoccus, Agrobacterium*, *Pseudomonas* and *Actinobacteria*, and a few Gram-positive bacteria like *Bacillus licheniformis*. Besides complete nitrogen removal, the promising application of heterotrophic aerobic denitrifiers along with nitrification and denitrification systems poses advantages over conventional systems with respect to cost-effectiveness and minimum sludge.

9.5 Conventional Methods for Nitrogen Removal from Wastewater

In general, there are two main methods of nitrogen removal from wastewater – physical-chemical treatments and biological treatments. Air-stripping, coagulation–flocculation, absorption, chemical oxidation and membrane separation are the main physical–chemical treatments used, whereas the biological treatments include all of the existing nitrogen removal processes, including the conventional nitrification and denitrification processes, nitritation–denitritation processes,[11] the endogenous denitritation process[12] and the anammox process.[13] The conventional methods or the possible pathways for the removal of nitrogen from wastewater are derived typically from the nitrogen cycle (Figure 9.2).

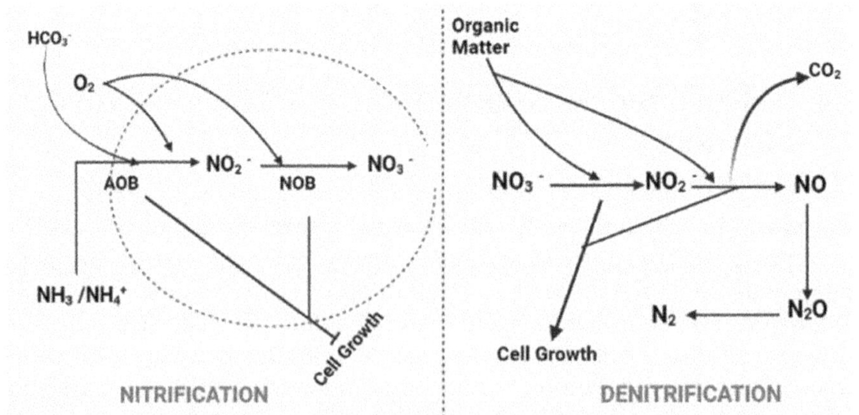

Figure 9.2 Conventional biological processes of nitrogen removal comprising mainly two processes, namely nitrification and denitrification.

9.6 Nitrification–Denitrification Process of Nitrogen Removal

The conventional nitrification-denitrification process[14] involves the conversion of NH_4^+–N into NO_3^-–N in aerobic conditions with further reduction to N_2 in the subsequent anoxic conditions. Nitrification occurs in two processes in the aerobic stage (Bove *et al.*, 2015), where the first stage involves the oxidation of NH_4^+–N into NO_2^-–N by ammonia oxidizing bacteria (AOB) and the second subsequent stage displays the conversion of NO_2^-–N into NO_3^-–N with the help of nitrite oxidizing bacteria (NOB). Following aeration, denitrification occurs involving the conversion of NO_3^-–N to N_2 during an anoxic stage with the aid of heterotrophic bacteria in the presence of organics. Ammonium or nitrite is mainly used as an energy source and electron acceptor.[15] More commonly, *Nitrosomonas* is mostly used in the oxidation of ammonium ions. Other bacteria involved in ammonium oxidation represent genera *Nitrosococcus*, *Nitrospira* and *Nitrosovibrio*. Nitrite oxidation involves *Nitrospira, Nitrococcus*, *Nitrosovibrio*, *Nitrospina* and *Nitrocystis* bacterial genera. *Nitrobacter* is most commonly used as a nitrite oxidizer.[16]

9.7 The Nitritation–Denitritation Process

The conversions of NO_2^-–N to NO_3^-–N and NO_3^-–N to NO_2^-–N lead to an increase in the consumption of more carbon sources as well as a high amount of energy. Denitrification could occur only with nitrite and not with nitrate only if ammonia, which is oxidized to nitrite (also called partial nitrification) would not be further oxidized to nitrate, meaning that nitrification would have been controlled at a much earlier stage. The most perilous step in the nitritation–denitritation process is to achieve stable nitritation.[17] It is carried out to promote the out-selection of the nitrite oxidizing bacteria in an intermittently aerated bioreactor where the intermittent aeration is controlled using a strategy based on effluent ammonia and the combination of nitrate and nitrite concentrations. In addition, a novel biological nitrogen removal system has been developed based on nitritation coupled with thiosulfate-driven denitritation to achieve a high rate of nitrogen removal and low sludge production.[18]

9.8 The Nitritation–Endogenous Denitritation Process

The nitritation–endogenous denitritation process is an important and novel process that completely relies on intracellular polymers, and in comparison to the conventional nitrification–denitrification process where external organic materials need to be completely added to remove nitrogen, this process does not require any additional external carbon sources for nitrogen removal. Instead, this process makes use of organic waste materials, thereby decreasing the cost and complexity of wastewater treatment. This process can be used to effectively treat young and medium landfill leachate containing biodegradable organics with a high COD/N ratio.

9.9 Advanced Approaches for Biological Nitrogen Recovery

Apart from the conventional technologies used for nitrogen removal from wastewater, new technologies have been developed recently or are under development for the removal of nitrogen. The technologies developed or the main processes for nitrogen elimination include partial nitritation, nitrification–denitrification, the Annamox method and their combinations like CANON and SHARON.[19,20] The combination of two technologies, *i.e.* partial nitritation with annamox (anaerobic ammonium oxidation) for the removal of nutrients, has turned out to be effective and much more beneficial rather than conventional processes due to the fact that it does not require any additional external carbon sources; instead, it requires low oxygen and energy. Additionally, n-DAMO coupling to the annamox process could provide an opportunity for simultaneous nitrogen removal[21] and methane emission mitigation from wastewater. Researchers have developed an operation strategy on the basis of the interactions among the n-DAMO and annamox microorganisms in a MBfR (membrane biofilm reactor)[22] for the removal of nitrogen from landfill leachate and wastewater.[23] Some of the other nitrogen removal processes include the OLAND (oxygen limited autotrophic nitrification and denitrification) process, NO_x process, CANDO (coupled aerobic–anoxic nitrous decomposition operation) process and n-DAMO (nitrate/nitrite-dependent anaerobic methane oxidation) process. The following section gives a brief description of all of the advanced methods for the removal of nitrogen.

9.10 The Anammox Method

The anaerobic ammonium oxidation (annamox) method was perceived within a fluidized bed bioreactor.[24] It is a well-known low-energy consuming eco-friendly process for the treatment of nitrogen rich waste-water (shown in Figure 9.3A). It outperforms the conventional methods of nitrification and denitrification technology since the nitrogen elimination rate is higher with less operational cost. The anammox system has gained immense attention since its discovery in the early 1990s and has been implemented in quite a few wastewater technologies, which has resulted in higher rates of nitrogen removal and lower energy requirements, thereby saving operational costs as compared to the conventional wastewater technologies. It has ecologically benefitted wastewater treatment. Anammox bacteria belonging to the Planctomycetes family contain a membrane-bound organelle that converts

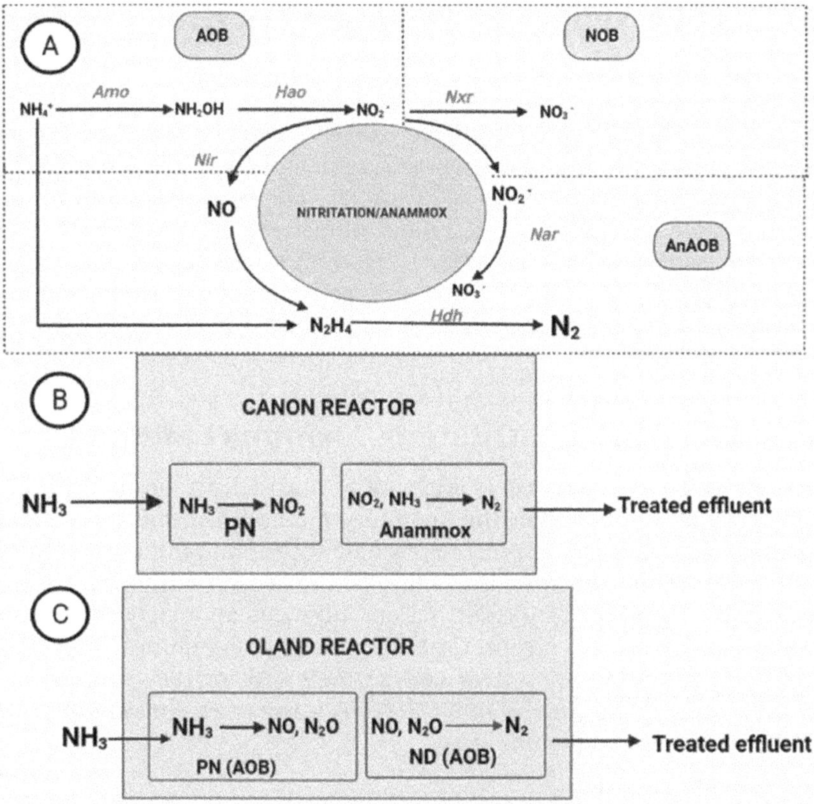

Figure 9.3 Three processes (A–C) showing the advanced methods of nitrogen removal from wastewater, namely the anammox process, canon process and OLAND process, respectively (PN – Partial Nitrification, AOB – Ammonia Oxidizing Bacteria, ND – Nitrifier Denitrification, DN – Denitrification, NOB – Nitrite Oxidizing Bacteria).

both the nitrite (NO_2^-) and ammonium ions (NH_4^+) to form dinitrogen gas (N_2) *via* nitrite-oxide (NO) and extremely energy-rich hydrazine (N_2H_4) as intermediates.

The water treatment process is a two-step process comprising a side stream and mainstream process that are segregated based on influent sources and bacterial involvement. The mainstream process involves the decoupling of carbon and nitrogen removal and helps in maximising energy recovery through a carbon-concentrating pre-treatment process channelling more carbon to an anaerobic digester for biogas production. The mainstream anammox process starts with two consecutive steps. Firstly, ammonium is primarily converted to nitrite facilitated by ammonium oxidizing bacteria (AOB), known as partial nitritation (PN), and in the second step the remaining ammonium reacts with nitrite to produce dinitrogen gas supported by nitrite oxidizing bacteria (NOB).

9.11 Ecophysiology of Anammox Bacteria

The known bacteria with the capacity for anaerobic ammonium oxidation are limited to a group of micro-organisms belonging to the phylum *Planctomycetales*. To date, the anammox bacteria belong to five *Candidatus* genera, which are monophyletic and branch off deep inside the Planctomycetes.[25–27] The lineages of descent, *Brocadia, Kuenenia*, *Anammoxoglobus* and *Jettenia*, share the same level of metabolism and ultrastructure. Anammox bacteria are omnipresent in man-made and natural oxygen-limited ecosystems. They are usually cultivated from the soil and activated sludge and have been confirmed as a kind of chemoautotrophic bacteria, thereby using carbon as the energy source.[28]

9.12 The Partial Nitritation–Anammox Method

Nitrogen, which is generated in wastewater and rich in ammonium compounds, can be eliminated by the Anammox method. The substance should be pre-oxidized such that approximately half of the ammonium is converted to nitrite before the substrate is fed into the anammox system, which indicates that the method requires nitritation. Advanced approaches like partial SHARON (Single Reactor system for High activity Ammonium Removal over Nitrite) and the SHARON system can be preferred for the elimination of ammonia through the nitrite route and for wastewater containing a high amount of concentrated ammonia.

9.13 Side-stream Partial Nitritation/Anammox

This is an energy-saving modification to the earlier nitritation–denitritation process and combining partial nitritation with the anammox process has impacted largely on the nitrogen removal process. The partial nitritation/anammox process has been designed as a one or two step process where half of the ammonia is oxidized to nitrite by ammonia oxidizing bacteria (AOB)[29] and the resulting ammonium and nitrite rich stream is fed into the anammox reactor. Side-stream partial anammox is considered to be a low-energy process for nitrogen reduction, being recycled at the front of the treatment plant. However, due to high nitrogen concentrations, this side-stream process is suited for high recovery of nitrogen by physical/chemical approaches or microbial electrochemical cells and is mainly affected by high carbon content in the system.

9.14 Mainstream Partial Nitritation/Anammox

The high level of success in side-stream partial nitritation treatment and the dramatic savings in carbon content and aeration energy have led to interest from researchers in applying partial nitritation to mainstream treatment. A mainstream partial nitritation system[30] is quite challenging due to some key factors that affect the instability and inflexibility of effluent nitrogen concentrations. The lower temperature disproportionately favours nitrite oxidizing bacteria (NOB) relative to Anammox and Ammonium Oxidizing Bacteria (AOB). The COD/N ratio is higher leading to an excess of heterotrophic growth, and there is a longer start-up period, lower nitrogen concentrations and inconsistency in the loading rate of ammonia, which thereby restricts the Anammox and AOB growth.[31]

Environmental factors such as temperature, oxygen concentration, organic carbon content, nitrogen concentration and pH are the main factors influencing the mainstream process.

Investigations have been carried out at different scales of operation, including industrial or municipal wastewater treatment units, regarding the influence of environmental factors in each bioreactor whether in the mainstream or side-stream process, but conclusions couldn't be inferred from those investigations to understand the challenges posed by environmental factors.

Emerging research strategies have been developed for mainstream anammox, which include coupling the anammox process with either ammonium oxidizing archaea (AOA) or nitrate-dependent anaerobic methane-oxidizing (n-DAMO) archaea.[32] Ammonium oxidation to nitrite is done by AOA just like AOB, but they possess a higher affinity for ammonia and oxygen, which makes them better suited for mainstream PNA with low effluent ammonia concentrations. Apart from this, n-DAMO can reduce nitrate to nitrite with methane as an electron donor, offering a more consistent and reliable nitrite supply to anammox than AOB and thereby eliminating the requirement for NOB suppression. The sulfate reduction, autotrophic denitrification, nitrification integrated (SANI) process is another autotrophic ammonium removal process besides anammox where organic carbon is oxidized with sulfate as the electron acceptor instead of utilizing energy intensive aeration and the formed sulphide is used to reduce nitrate from a nitrification reactor to dinitrogen gas.[33]

9.15 The CANON Method

Completely autotrophic nitrogen removal over nitrite (CANON) is a cost-effective nitrogen removal process[34] that mainly relies on the cooperation/amalgamation of incomplete nitritation and Anammox working together within a lone reactor, as well as inhibition of the nitrite-oxidizing bacteria, and produces low amounts of sludge as compared to the conventional processes (as shown in Figure 9.3B). The main operational strategies include strict limitations on dissolved oxygen (DO) concentration in the reactor and the supplementing sufficient inorganic carbon in the influent. The two sequential reactions involved rely on the simultaneous interaction of two sets of autotrophic bacteria under oxygen-limited conditions.[35] Reports have suggested that the CANON process provides several advantages over the combination of partial nitritation and the anammox process, but is avoided for ammonium rich wastewater due to the difficulty it poses in maintaining the dissolved oxygen levels in large tanks and incomplete nitrogen removal. Despite all of the research to date, a better understanding of the oxygen-limited conditions for improving and optimizing the CANON process is required for effective nitrogen removal.

9.15.1 Associated Nitrogen Removal Processes

9.15.1.1 NO_x Process

The NO_x process involves the addition of trace concentrations of nitrogen oxides (NO or NO_2) within the wastewater to control the denitrification activity of *Nitrosomonas* bacteria. Under complete oxic conditions, the added NO_x works as a controller in nitrification and denitrification by *Nitrosomonas*-like bacteria.

9.15.1.2 CANDO Process for N_2O Production

Coupled Aerobic–Anoxic Nitrous Decomposition Operation (CANDO) is a three step process[36,37] that enables direct energy recovery for wastewater treatment with the subsequent generation of nitrous oxide (N_2O).[38] Nitrous oxide is a strong greenhouse gas formed during wastewater nitrogen removal processes. This process involves partial nitrification of NH_4^+ to NO_2^-, which is then anaerobically reduced to N_2O by denitrification, followed by the final conversion of N_2O to N_2 with the generation of energy. Reports suggest that about 60–80% of nitrite could be converted to nitrous oxide in the CANDO process and the specific nitrite oxide generation rate would be as high as 25.2 mg N per (g VSS h).

9.15.1.3 OLAND Method

Oxygen-limited autotrophic nitrification and denitrification (OLAND) is a biological nitrogen removal process coupled with partial nitrification and anaerobic ammonium oxidation. The conventional process for ammonium removal *via* two steps, aerobic nitrification and anaerobic denitrification, is challenged by a one-step process, namely the OLAND process (as shown in Figure 9.3C), where direct oxidation of ammonium takes place with the release of dinitrogen gas.[39,40] This autotrophic process consumes 63% less oxygen and 100% less reducing agent, yet has a higher efficiency of organic carbon removal than the conventional nitrification and denitrification method.[41] Applications of this one-step process with nitrifiers as a biocatalyst are limited due to the extremely low specific capacity of nitrogen removal.

9.15.1.4 n-DAMO Process

The nitrate/nitrite-dependent anaerobic methane oxidation (n-DAMO) process is a new sustainable wastewater treatment process operating without any requirement of external dosing of electron donors for denitrification. Rather, the methane is produced by the methanogens during the anaerobic digestion of disposal sewage sludge. n-DAMO in coupling to the anammox process provides a chance for simultaneous nitrogen removal performance of the anammox process using methane from anaerobic digestion.

9.16 Challenges and Future Prospects

The shift from chemical to biological methods of wastewater treatment is undoubtedly the need of the hour. The increasing toxicity posed by the chemical methods will further enhance the deleterious effect on the environment. Hence, a less toxic and more sustainable approach is needed for efficient nitrogen removal from wastewater plants. Critical factors like maintaining low capital and operating costs, low aquatic toxicity, operational flexibility, reduction in sludge production, reduction in filamentous growth, increase in sludge settling ability, improved sludge dewatering, and reduction in oxygen requirement offer challenges to stabilize the economical running and efficiency of the treatment plant. In recent years, several new biological nitrogen removal technologies (*i.e.* anammox and DAMO) have been developed that might help reduce costs and increase the efficacy of tertiary wastewater treatment significantly. Although many of the microorganisms involved in various biological treatment processes have been identified, there is limited information regarding their metabolisms and evolution patterns.

Although activated sludge systems are composed of large numbers of microorganisms that actively contribute to the removal of nutrients responsible for eutrophication of water resources, there is still a need for more nutrient-related research monitoring in order to achieve unpolluted wastewater.[42] The metabolic versatilities, adaptive mechanisms, and application potential in optimizing nitrogen removal in current and new bioreactors of these organisms are of great interest to be investigated in further studies. Furthermore, the symbiotic relationships between autotrophic nitrogen removal microorganisms and the dominant heterotrophic microorganisms deserve more attention. The roles of variables in shaping nitrogen-transforming microbial networks, such as the niche preference, alternative substrates, cellular energy requirements, and sensitivity to physicochemical factors, are a subject of great interest to provide a valuable reference for the establishment and operation of sewage treatment plants.

9.17 Conclusion

A considerable number of site-specific and laboratory studies have been conducted to evaluate the feasibility and performance of conventional methods for the elimination of organics and nutrients in wastewater. Various modified and/or altered conventional processes are operational based on the metabolism and microbiology involved for the elimination of nitrogen and phosphorous. The obtained results indicate nitrogen and phosphorous removal up to 60–70% using high-rate biological processes. However, ideal systems for efficient removal of nutrients for low to medium strength wastewater are yet to be devised and operational. Extensive work is still required to evaluate the possible causes of the poor performance or for possible up-gradation and/or modification of the existing treatment concept and/or pathway. In addition, multi-omics approaches like metagenomics along with microbiome investigation of wastewater treatment methods are required in order to augment knowledge related to genes implicated in wastewater treatment procedures based on distinct bacterial populations.

Conflicts of Interest

No potential conflict of interest is reported by the authors.

Abbreviations

WWTPs	Wastewater treatment plants
AOB	Ammonia oxidizing bacteria
NOB	Nitrite oxidizing bacteria
CANON	Completely autotrophic nitrogen removal over nitrite
Anammox	Anaerobic ammonium oxidation
SHARON	Single reactor system for high activity ammonium removal over nitrite
DNRA	Dissimilatory nitrite reduction to ammonium
n-DAMO	Nitrite/nitrate dependent anaerobic methane oxidation
OLAND	Oxygen-limited autotrophic nitrification and denitrification
CANDO	Coupled aerobic–anoxic nitrous decomposition operation
COD	Chemical oxygen demand
AMO	Ammonia monooxygenase
HAO	Hydroxylamine oxidoreductase
NFB	Nitrogen-fixing bacteria
BNF	Biological nitrogen fixation

Acknowledgements

The authors are thankful for the assistance and support extended by the Director, School of Biotechnology, Kalinga Institute of Industrial Technology (KIIT) Deemed-to-be-University, Odisha, for extending the infrastructure facilities.

References

1. D. E. Holmes, Y. Dang and J. A. Smith, Nitrogen cycling during wastewater treatment, *Adv. Appl. Microbiol.*, 2019, **106**, 113–192.
2. G. Gulis, M. Czompolyova and J. R. Cerhan, An ecologic study of nitrate in municipal drinking water and cancer incidence in Trnava District, Slovakia, *Environ. Res.*, 2002, **88**(3), 182–187.
3. O. Akpor and M. Muchie, Remediation of heavy metals in drinking water and wastewater treatment systems: processes and applications, *Int. J. Phys. Sci.*, 2010, **5**(12), 1807–1817.
4. M. K. Winkler and L. Straka, New directions in biological nitrogen removal and recovery from wastewater, *Curr. Opin. Biotechnol*, 2019, **57**, 50–55.
5. B. Swaminathan and K. Sukalac, *Technology transfer and mitigation of climate change: The fertilizer industry perspective,* Paper presented at the IPCC Expert Meeting on Industrial Technology Development, Transfer and Diffusion, Tokyo, Japan, 2004.
6. M. Ali and S. Okabe, Anammox-based technologies for nitrogen removal: advances in process start-up and remaining issues, *Chemosphere*, 2015, **141**, 144–153.
7. B. Ma, S. Wang, S. Cao, Y. Miao, F. Jia, R. Du and Y. Peng, Biological nitrogen removal from sewage via anammox: recent advances, *Bioresour. Technol.*, 2016, **200**, 981–990.
8. H. I. Gomes, Phytoremediation for bioenergy: challenges and opportunities, *Environ. Technol. Rev.*, 2012, **1**(1), 59–66.
9. R. Dixon and D. Kahn, Genetic regulation of biological nitrogen fixation, *Nat. Rev. Microbiol.*, 2004, **2**(8), 621–631.
10. G. Guo, Y. Wang, T. Hao, D. Wu and G.-H. Chen, Enzymatic nitrous oxide emissions from wastewater treatment, *Front. Environ. Sci. Eng.*, 2018, **12**(1), 1–12.
11. Q. Yang, Y. Peng, X. Liu, W. Zeng, T. Mino and H. Satoh, Nitrogen removal via nitrite from municipal wastewater at low temperatures using real-time control to optimize nitrifying communities, *Environ. Sci. Technol.*, 2007, **41**(23), 8159–8164.
12. K. Wang, S. Wang, R. Zhu, L. Miao and Y. Peng, Advanced nitrogen removal from landfill leachate without addition of external carbon using a novel system coupling ASBR and modified SBR, *Bioresour. Technol.*, 2013, **134**, 212–218.
13. H. Li, S. Zhou, W. Ma, P. Huang, G. Huang, Y. Qin and H. Ouyang, Long-term performance and microbial ecology of a two-stage PN–ANAMMOX process treating mature landfill leachate, *Bioresour. Technol.*, 2014, **159**, 404–411.
14. J. Ibrahimpašić, T. Landeka Dragičević, M. Zanoški Hren, V. Bačun-Družina, M. Čurlin and I. Vinković Vrček, Nitrogen removal from municipal landfill leachate, *Chem. Biochem. Eng. Q.*, 2010, **24**(4), 453–458.

15. G. Tchobanoglus, F. Burton and H. D. Stensel, Wastewater engineering: Treatment and reuse, *J. Am. Water Works Assoc.*, 2003, **95**(5), 201.
16. Y. Cao, M. van Loosdrecht and G. T. Daigger, Mainstream partial nitritation–anammox in municipal wastewater treatment: status, bottlenecks, and further studies, *Appl. Microbiol. Biotechnol.*, 2017, **101**(4), 1365–1383.
17. P. Regmi, B. Holgate, D. Fredericks, M. W. Miller, B. Wett, S. Murthy and C. B. Bott, Optimization of a mainstream nitritation–denitritation process and anammox polishing, *Water Sci. Technol.*, 2015, **72**(4), 632–642.
18. J. Qian, J. Zhou, Z. Zhang, R. Liu and Q. Wang, Biological nitrogen removal through nitritation coupled with thiosulfate-driven denitritation, *Sci. Rep.*, 2016, **6**(1), 1–10.
19. G. T. Daigger, Oxygen and carbon requirements for biological nitrogen removal processes accomplishing nitrification, nitritation, and anammox, *Water Environ. Res.*, 2014, **86**(3), 204–209.
20. W.-B. Nie, G.-J. Xie, J. Ding, L. Peng, Y. Lu, X. Tan and J. Meng, Operation strategies of n-DAMO and anammox process based on microbial interactions for high rate nitrogen removal from landfill leachate, *Environ. Int.*, 2020, **139**, 105596.
21. K. J. Martin and R. Nerenberg, The membrane biofilm reactor (MBfR) for water and wastewater treatment: principles, applications, and recent developments, *Bioresour. Technol.*, 2012, **122**, 83–94.
22. L. Miao, G. Yang, T. Tao and Y. Peng, Recent advances in nitrogen removal from landfill leachate using biological treatments–A review, *J. Environ. Manage.*, 2019, **235**, 178–185.
23. A. Mulder, A. A. Graaf, L. A. Robertson and J. G. Kuenen, Anaerobic ammonium oxidation discovered in a denitrifying fluidized bed reactor, *FEMS Microbiol. Ecol.*, 1995, **16**(3), 177–183.
24. I. Schmidt, O. Sliekers, M. Schmid, E. Bock, J. Fuerst, J. G. Kuenen and M. Strous, New concepts of microbial treatment processes for the nitrogen removal in wastewater, *FEMS Microbiol. Rev.*, 2003, **27**(4), 481–492.
25. S. Khramenkov, M. Kozlov, M. Kevbrina, A. Dorofeev, E. Kazakova, V. Grachev and Y. A. Nikolaev, A novel bacterium carrying out anaerobic ammonium oxidation in a reactor for biological treatment of the filtrate of wastewater fermented sludge, *Microbiology*, 2013, **82**(5), 628–636.
26. M. Kumar and J.-G. Lin, Co-existence of anammox and denitrification for simultaneous nitrogen and carbon removal—strategies and issues, *J. Hazard. Mater.*, 2010, **178**(1–3), 1–9.
27. N. Shivaraman and G. Shivaraman, Anammox-A novel microbial process for ammonium removal, *Curr. Sci.*, 2003, **84**(12), 1507–1508.
28. Y. Kimura, K. Isaka and F. Kazama, Effects of inorganic carbon limitation on anaerobic ammonium oxidation (anammox) activity, *Bioresour. Technol.*, 2011, **102**(6), 4390–4394.
29. S. Lackner, E. M. Gilbert, S. E. Vlaeminck, A. Joss, H. Horn and M. C. van Loosdrecht, Full-scale partial nitritation/anammox experiences–an application survey, *Water Res.*, 2014, **55**, 292–303.

30. M. Laureni, P. Falås, O. Robin, A. Wick, D. G. Weissbrodt, J. L. Nielsen and A. Joss, Mainstream partial nitritation and anammox: long-term process stability and effluent quality at low temperatures, *Water Res.*, 2016, **101**, 628–639.
31. M. S. Jetten, M. Schmid, I. Schmidt, M. Wubben, U. Van Dongen, W. Abma and L. Ottosen, Improved nitrogen removal by application of new nitrogen-cycle bacteria, *Rev. Environ. Sci. Biotechnol.*, 2002, **1**(1), 51–63.
32. T. Limpiyakorn, M. Fürhacker, R. Haberl, T. Chodanon, P. Srithep and P. Sonthiphand, amoA-encoding archaea in wastewater treatment plants: a review, *Appl. Microbiol. Biotechnol.*, 2013, **97**(4), 1425–1439.
33. D. Wu, G. A. Ekama, H.-K. Chui, B. Wang, Y.-X. Cui, T.-W. Hao and G.-H. Chen, Large-scale demonstration of the sulfate reduction autotrophic denitrification nitrification integrated (SANI®) process in saline sewage treatment, *Water Res.*, 2016, **100**, 496–507.
34. T. Khin and A. P. Annachhatre, Novel microbial nitrogen removal processes, *Biotechnol. Adv.*, 2004, **22**(7), 519–532.
35. J. G. Kuenen, Anammox bacteria: from discovery to application, *Nat. Rev. Microbiol.*, 2008, **6**(4), 320–326.
36. Z. Wang, S.-G. Woo, Y. Yao, H.-H. Cheng, Y.-J. Wu and C. S. Criddle, Nitrogen removal as nitrous oxide for energy recovery: Increased process stability and high nitrous yields at short hydraulic residence times, *Water Res.*, 2020, **173**, 115575.
37. Z. Zhang, Y. Zhang and Y. Chen, Recent advances in partial denitrification in biological nitrogen removal: From enrichment to application, *Bioresour. Technol.*, 2020, **298**, 122444.
38. S. S. Shalini and K. Joseph, Nitrogen management in landfill leachate: Application of SHARON, ANAMMOX and combined SHARON–ANAMMOX process, *Waste Manage.*, 2012, **32**(12), 2385–2400.
39. L. Kuai and W. Verstraete, Ammonium removal by the oxygen-limited autotrophic nitrification–denitrification system, *Appl. Environ. Microbiol.*, 1998, **64**(11), 4500–4506.
40. S. Vlaeminck, H. De Clippeleir and W. Verstraete, Microbial resource management of one-stage partial nitritation/anammox, *Microb. Biotechnol.*, 2012, **5**(3), 433–448.
41. K. Windey, B. I. De and W. Verstraete, Oxygen-limited autotrophic nitrification–denitrification (OLAND) in a rotating biological contactor treating high-salinity wastewater, *Water Res.*, 2005, **39**(18), 4512–4520.
42. O. Akpor and M. Muchie, Bioremediation of polluted wastewater influent: Phosphorus and nitrogen removal, *Sci. Res. Essays*, 2010, **5**(21), 3222–3230.

CHAPTER 10

Nitrite-oxidizing Bacteria: Cultivation, Growth Physiology, and Chemotaxonomy

POOJA M. PATIL,[a] ARUN KUMAR PARTHASARATHY,[b] ABHIJEET R. MATKAR,[c] PRANJALI MAHAMUNI-BADIGER,[d] SANGCHUL HWANG,[e] RANJIT GURAV*[e] AND MARUTI J. DHANAVADE*[f]

[a] Department of Environment Management, Chhatrapati Shahu Institute of Business Education and Research, Kolhapur, India; [b] Department of Microbiology, D.Y. Patil University, Kolhapur, India; [c] Department of Mechanical Engineering, D.Y. Patil College of Engineering and Technology, Kolhapur, India; [d] Department of Microbiology, Rayat Shikshan Sanstha's, S.M. Joshi College, Hadapsar, Pune, India; [e] Ingram School of Engineering, Texas State University, San Marcos, Texas, USA; [f] Department of Microbiology, Bharati Vidyapeeth's Dr. Patangrao Kadam Mahavidyalaya, Sangli, India

*Emails: rnjtgurav@gmail.com; marutijd@gmail.com

10.1 Introduction

Nitrification is the two-step process in which the oxidation of ammonia to nitrate is performed by chemolithoautotrophic microorganisms. The first step of nitrification is the oxidation of ammonia to nitrite by ammonia-oxidizing bacteria (AOB), and the second step is the oxidation of nitrite to nitrate by nitrite-oxidizing bacteria (NOB). Nitrogen is one of the important water pollutants in wastewater generated by agricultural

Ammonia Oxidizing Bacteria: Applications in Industrial Wastewater Treatment
Edited by Maulin P. Shah

Published by the Royal Society of Chemistry, www.rsc.org

and industrial activities. A large amount of nitrogen reaches different water bodies in the form of nitrates, nitrites, urea, and ammonia, causing detriment to aquatic life, and environmental changes like eutrophication.

$$NH_3 + O_2 \rightarrow NO_2^- + 3H^+ + 2e^-$$

$$NO_2^- + H_2O \rightarrow NO_3^- + 2H^+ + 2e^-$$

Any treatment plant that discharges nitrogen into water bodies may have a total nitrogen limit of 3 $mg\,L^{-1}$ or a unionized ammonia limit of 0.2 $mg\,L^{-1}$. To meet the above limits of nitrogen, treatment plants should remove or convert nitrogenous compounds to less toxic forms in addition to the removal of total suspended solids (TSS) and carbonaceous biochemical oxygen demand (CBOD). Domestic sewage may contain 20 to 40 $mg\,L^{-1}$ of nitrogen in the form of ammonia. So, industrial water should be processed to remove nitrogenous waste before discharging into that reservoir.

Industrial wastewater has been processed by physical, chemical, and biological methods to remove heavy metals, toxic chemicals, and other impurities from the wastewater.[1–8] Each of these methods has some limitations in terms of cost, applicability, and efficacy. Microorganisms are an effective biological asset to completely remove nitrogen, which is generated by wastewater from different sources. Microbial bioremediation of nitrogenous contaminated industrial wastewater can be carried out through the denitrifying and native nitrifying bacteria that use inorganic nitrogen or ammonium as the sole nitrogen source produced by various human activities. NOBs are autotrophs, thus using inorganic carbon sources like carbon dioxide, carbonate ions, and simple inorganic nitrogen for the metabolic synthesis of organic molecules. These autotrophs grow very slowly compared to heterotrophs (one cell dividing into two takes 8 h while heterotrophs take 20 min). In addition, NOBs are very sensitive to temperature, pH, and chemicals. The identification of such bacteria is more important, particularly for determining their intrinsic potential for use in the bioremediation of wastewater. Although NOB species are efficient to do nitrification, there are still a few disadvantages as their slow-growing nature can interfere with the formulation of microbial products for the bioremediation process. However, enrichment techniques can be applied to test the efficiency of NOB species in nitrification as well as understand the microbial diversity of NOBs. In this book chapter, all the aspects related to NOBs like overview, history, different NOB communities, growth physiology, cultivation, identification, and NOB chemotaxonomy have been well explained. A thorough understanding of these points could be very useful to expand the existing knowledge and search for novel species with distinct characteristics in the near future.

10.2 Overview and History of NOBs

At present, novel NOB species are rarely isolated in various culture collections. This limited number of new NOB isolates has created speculation that most of the species may be present in uncultured form (Spieck and Bock, 2005). NOBs are difficult to culture because they are slow growers and remain active in standard preservation procedures such as freeze-drying. When the NOB growth parameters are adapted closer to natural habitats, the existing diversity in collecting these NOB strains can be achieved. It is known that nitrite rarely accumulates in nature, and overexposure to this toxic compound may lead to a hazardous effect. Most species from the genus *Nitrobacter* were isolated at an initial nitrite concentration of 29 mM. However, at low nitrite concentrations like 1 mM, marine strains like *Nitrococcus, Nitrospina* and *Nitrospira* were obtained. As such, the cultivation of marine strains (*Nitrospira*, *Nitrococcus*, and *Nitrospina*) required a lower concentration of 1 mM nitrite.[9] As per an earlier protocol, the nitrite concentrations were kept at 0.05 mM, which was very low. But it was determined that varying soil populations of NOB have different nitrite requirements.[10] The slight modification of using 2.9 mM instead of 29 mM nitrite has resulted in the growth and enrichment of *Nitrospira* species selectively in activated sludge.[11] A further 10-fold dilution of nitrite to 0.29 mM leads to the growth of novel *Nitrotoga* from permafrost-affected soils.[12] Later findings revealed that low incubation temperature (10 °C) is also a key factor for the growth of cold-loving NOB present in moderate-climate wastewater treatment units.[13] The NOB inhabitants in hot springs need 60 °C temperatures for their isolation.[14] Overall, from this earlier research, it is well-studied that the enrichment of NOBs from diverse natural sources and habitats is possible. Hence, it is easily speculated that there are still miles to go in understanding the whole diversity of NOBs. Some of the points discussed below in this chapter could be very useful to rediscover and study one of the important but overlooked microbial communities, like NOBs which carry out a vital step in nitrification.

10.3 The Nitrogen Cycle and NOBs

It is well known that nitrogen is very essential for all living species for their survival. Microorganisms play a very crucial role in the biogeochemical nitrogen cycle (Figure 10.1). These small creatures mediate the conversions of nitrogenous compounds, like N_2 fixation into organic molecules as well as recycling nitrogen from decaying biomass into the atmosphere. The sequential oxidation of ammonia *via* nitrite to nitrate is an important step in the nitrogen cycle. Earlier, the great work of scientist Sergei Winogradsky to isolate the first chemolithoautotrophic bacteria which grew by nitrification using ammonia and nitrite as a sole energy source was the basis of today's knowledge in this field.[15] During his pioneer work, Winogradsky discovered two nitrification steps that were catalyzed by different bacteria namely AOB

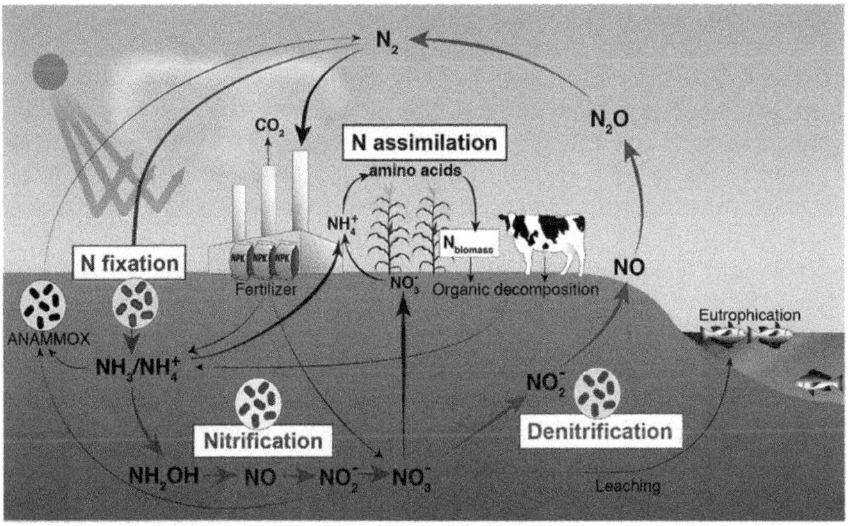

Figure 10.1 A significant transition of the nitrogen cycle. Schematic diagram and the principal steps in the nitrogen cycle. The atmospheric nitrogen is fixed in the soil by nitrogen-fixing bacteria, then nitrification by nitrifying bacteria, after assimilation occurs and the last step is denitrification. Reproduced from ref. 59 with permission from the Royal Society of Chemistry.

and NOB. The cooperation of these two important microorganisms is very essential to achieve complete nitrification. During his study he found that both these microorganisms were slow in growth; hence, to cultivate these microbes in the lab one should have perseverance.[15] In the succeeding year after the work of Winogradsky, nitrification research progressed well to determine the biochemistry of NOB, especially the genera *Nitrosomonas* (AOB) and *Nitrobacter* (NOB). This research was mainly driven by the thrust as well as curiosity to know the importance of the nitrification process. It is known in agriculture that the conversion of fertilizer ammonium into nitrate causes considerable nitrogen loss from the soil, which decreases its fertility. However, the nitrification process proved to be advantageous as an important step in treating biological wastewater to remove excess nitrogen present in sewage. Several decades of research were focused on AOB and their nitrification ability. The main reason behind this research direction is that the ammonia oxidation process is supposed to be the rate-limiting nitrification step. During that period, the research on ammonia oxidation was shooting up with the discovery of ammonia-oxidizing archaea (AOA). In contrast, there were difficulties growing NOB in the laboratory as compared to AOA or AOB. Usually, nitrates do not accumulate in the natural environment; hence, their oxidation does not gain attention. Additionally, NOB are well known as obligate chemolithoautotrophs having a very partial range of physiological characteristics. Hence, there is a lower probability to discover new NOB with unique physiology. Current research is lacking to find newer nitrogen-cycling microorganisms and their importance. Whether the fixed nitrogen persists in an ecosystem or gets lost in the atmosphere can be determined by the fate of nitrate. NOB convert nitrite to nitrate and stop the nitrogen loss. This nitrate is a good and utilizable nitrogen source used by many plants as well as microorganisms.[16] Hence, this important function of NOB highlighted that they could play a vital role in the nitrogen cycle. It is known that nitrite is lethal to eukaryotic organisms and it can inhibit bacterial growth.[17] NOB activity in industrial wastewater treatment plants (WWTPs) is found to be unstable and if nitrite from WWTPs leaks into natural water bodies it can damage the ecosystem drastically. The ecological importance of NOB pushed us to acquire new insights and gain good knowledge about the biology of NOB. At present, improving research efforts on NOB using metagenomics and other metaomics approaches could be useful for better understanding. Studies like single-cell isotope labelling along with some other genomic analytical techniques supporting refined cultivation methods can offer the best chance to study NOB.[18] There are chemolithoautotrophic NOB as well as phototrophic NOB, which have to be studied in detail due to their tremendous potential.

10.4 Habitats of NOB

NOB are distributed widely and have evolved a different diversity of microbial populations. They are adapted to variable environmental conditions and detected in geothermal springs,[12] permafrost soils,[14] alkaline soils, and sediments,[19] and play a major role in rock erosion. Six genera of aerobic chemolithoautotrophic NOB are available to date. NOB are usually associated with AOB, providing a nitrogen energy source to the bacteria. Multiplication of NOB in sewage plays a major role in removing nitrogenous waste from industrial wastewater.

NOB grow at a wide range of pH ranges up to 10 and higher salinities.[19] NOB are obligate aerobes, but the growth of NOB is observed in anaerobic environments such as the deep bottom mud of anaerobic wastewater reservoirs and with significant concentrations of sulfides. Apart from these, NOBs are also isolated from marine anaerobic environments from regions located at a depth of around 8 cm.[20]

10.5 Biofilm Formation in Rhizobia and Enumeration Methods for NOB

Biofilm formation allows protected growth under unfavourable environmental conditions like desiccation, UV exposure, predation, and antibiotics. The plant root zonal area possesses large numbers of biological and ecological processes, which is known as the Rhizosphere. It provides favorable conditions for biofilm formation, like the availability of moisture, and nutrients.

Biofilm formation on plants is symbiotic and pathogenic also. Under the deficiency of nitrogen, the *Rhizobia* bacterium associates with the roots of leguminous plants and forms a symbiotic association with nodules. Inside these nodules, atmospheric nitrogen gets converted into ammonia by a reduction reaction, which is used by the plants and the bacteria get carbohydrates from the plants. This process is followed by a signalling mechanism between legumes and *Rhizobia*.[21] The plant exudes flavonoids, which are sensed by the bacteria, and they induce the activation of *nod* genes with *nod* factors. At the time of this symbiosis, the bacteria attach and colonize the roots by the special structure termed 'infection threads'. At the time of nodule and biofilm formation, exopolysaccharides, lipopolysaccharides, flagella, signalling molecules, and capsular polysaccharides play a key role in the attachment of bacteria to the surface of the nodules where the bacteria can divide, and the community forms a three-dimensional structure in which channels are present. The bacterial cells can disperse from the biofilm, which occurs in the biofilm formed on the host plant root. According to various studies, biofilm formation helps *Rhizobia* under unfavourable environmental conditions.[22]

The microsensor and molecular techniques are combinedly use to study the microbial ecology of biofilms; also, these methods are particularly applied to finding the nitrifying bacterial sites. Earlier studies have reported the distribution of nitrifying bacteria of various genera like *Nitrosospira*, *Nitrosomonas*, *Nitrospina*, and *Nitrobacter* in biofilms.[22] In membrane-bound biofilms, the pH, oxygen, nitrate, and nitrite were easily determined by biosensors, whereas the NOB populations were quantified by FISH (Fluorescent *in situ* hybridization) and confocal laser scanning microscopy methods.

10.6 Nitrogen Fixation by NOB in Plants

Microbial associations with plants are significant and play a key role in assimilating plant nutrients, growth, and protection. NOB form efficient and well-defined associations in legumes. In a symbiotic association, the utilization of nitrogen in the form of nitrate, ammonium, and organic nitrogen takes place to fix nitrogen. Then the host plant absorbs and utilizes the ammonium formed by nitrogen-fixing endophytes. In this process, organic carbon, amino compounds, and mineral elements are provided by the plant host.

10.6.1 Symbiotic Nitrogen Fixation by Legume and Non-legume Nodules

It is widely reported that legumes can provide an ample amount of nitrogen in natural and cultivated ecosystems. The most common legumes used by human beings are Chickpeas (*Cicer arietinum*), Peanuts (*Arachis hypogea*), Pigeon peas (*Cajanus cajan*), Cowpeas (*Vigna unguiculata*), Lentils

(*Lens esculenta*), *etc.*[23] These legumes are highly efficient for nitrogen fixation and are mainly used for intercropping or crop rotation purposes. Symbiotic association between the soybean root system and *Rhizobia* fixes a large amount of nitrogen that plays a significant role in the growth, development, and maturity stages. Some legumes cultivated under a wide range of climatic spectrum possess high potential to supply enough nitrogen in the soil and in turn, give more yield. Most of the forage legumes like Alfalfa (*Medicago sativa*), Subterranean clover (*Trifolium subterraneum*), Red clover (*Trifolium pratense*), and White clover (*Trifolium repens*) are widely distributed in hot regions of the earth.[24] Several plant species which are economically important crops and some that can produce biofuel during nitrogen fixation are considered potential sources to develop soil fertility. For example, *Pongamia pinnata* is a medicinal and green manure plant that can form a well-defined symbiotic association with several strains like *Bradyrhizobium* and *Rhizobium*. This plant can fix nitrogen potentially and the amount of acetylene is reduced by nitrogenase enzymes, which are encoded by the bacteria.[25]

In non-symbiotic systems, bacteria fix nitrogen by utilizing carbon and energy from the environment. Fixed nitrogen is released mostly after the lysis of cells. In the Rhizosphere, associative systems between bacteria can fix nitrogen by utilizing carbohydrates from the environment, which are released from the roots, and degradation of organic material in the soil. Nitrogen fixation in non-leguminous plants has great agronomic significance. For example, sweet potato (*Ipomoea batatas*), *Bradyrhizobium,* and *Azorhizobium* species are present abundantly in nitrogen-fixing sugarcane stems, sweet potato stems, and tubers. A well-organized association between *Azospirillum brasilense* and the host plant *Setaria italica* can fix a huge amount of oxygen. Some more examples are the association between *Azoarcus* sp. *BH72*, *Kallar* grass, *Klebsiella pneumonia*, and wheat.

Maximum nitrogenase activities are present in the paddy rice-soil system as compared with paddy soil without rice plants. Nitrogen fixation in rice fields is performed by photosynthetic Cyanobacteria and heterotrophic Diazotrophs by utilizing root secretions. Aerial roots in maize play a significant role in increasing the nutrient, water uptake, and gaseous exchange between plant tissue and the atmosphere. Switchgrass (*Panicum virgatum* L.) can absorb freshly fixed nitrogen into its tissues. Different communities like *Burkholderia* and *Ralstonia taiwanensis* are predominantly present in the tissues of switchgrass.

The understanding of the detailed mechanisms responsible for plant-microbe association can help in transferring these genes in non-leguminous plants to enhance the fertility of the soil. Different studies of evolutionary genomics state that fewer genetic elements are required to transfer nitrogen-fixing capacity to non-legume plants. However, the transfer of nitrogenase genes to non-legume plants is difficult because of the complex structure of nitrogenase biosynthesis and the high sensitivity of the enzyme to oxygen.[26]

10.7 Insights into the Nitrite-oxidizing Bacterial Community

NOBs are Gram-negative bacteria that gain energy by converting nitrite to nitrate. The genus *Nitrobacter* was the first described nitrate oxidizer within the family *Alphaproteobacteria*.[15] The *Nitrobacter* genus is easy to cultivate when compared to other NOB and it requires a higher concentration of nitrites to be isolated in laboratories. So far, 4 species have been described, namely:

1. Nitrobacter winogradskyi isolated from soil.
2. Nitrobacter hamburgensis isolated from soil.[27]
3. Nitrobacter vulgaris isolated from various environments like freshwater, stones, and sewage.[28]
4. Nitrobacter alkalicus isolated from Soda Lake and Soda Lake sediments at pH 10.[19]

Nitrobacter species encode the gene subunits of the nitrite oxidoreductase enzymes, which are the key enzymes responsible for nitrite oxidation.[19] Genus *Nitrococcus* and *Nitrospinae* belong to the family *Gammaproteobacteria*. *Nitrococcus mobilis* originated from Pacific Ocean surface water,[29] whereas *Nitrospina gracilis* was isolated from Atlantic Ocean surface water.[9] Besides, *Nitrospina watsonii* was isolated from the marine environment.[30] The *Nitrospina* genus is isolated from most ocean water environments, like open oxygenated ocean water, marine sediments, anoxic marine sediments, and oxygen minimum zones.[31] Only a few species from the genus *Nitrococcus* have been identified to date and *Nitrococcus* is identified in ocean water and not seen in non-marine habitats; however, a few studies have identified *Nitrococcus* in oxygen minimum zone waters. The nitrite oxidation activities of *Nitrococcus* are limited to oxygen settings and they were measured at a higher rate of nitrite oxidation.[31]

The *Nitrospira* bacteria initially received less attention, but after identifying *Nitrospira* as NOB species in aquarium filters in 1998, they were the most abundant NOB in the nitrifying process. The genus *Nitrospira* is the most widely distributed NOB group. *Nitrospira marina* was the first isolated bacteria from the genus *Nitrospira* from an ocean water source.[32] Later after several years, *Nitrospira moscoviensis* was isolated from urban heating water.[33] The wide distribution of the *Nitrospira* genus was further verified by the culture of *Nitrospira defluvii* (highly enriched culture) from WWTP activated sludge.[34] In 2008, *Nitrospira bockiana* was isolated from geothermal springs,[35] and it is highly thermophilic and was isolated from the marine source Ecomares.[36] In 2013, two strains ND1 and *N. japonica* were isolated from activated sludge.[18,37] Based on 16S rRNA analysis, the wide distribution of *Nitrospira* bacteria was detected in different sources like sediments, different soils, freshwater, subsurface liquids, cave biofilms, hot springs, marine sponges, coastal water, recirculation aquaculture systems, *etc.* The genus *Nitrospira* has six phylogenetic lineages based on the 16S rRNA analysis (Figure 10.2 and Table 10.1).[38]

A cold strain of NOB from permafrost-affected soil was isolated and named as *Candidatus Nitrotoga Arctica* which belongs to beta-proteobacterium.[12] Nitrite oxidization is observed at temperatures between 4 °C and 22 °C.[12] Based on the 16S rRNA sequences, closely related *Nitrotoga*-like bacteria were observed in different samples like river biofilms, lake sediments, cave biofilms, bioreactors, water filters, *etc.*[39] In 2012, *Nitrolancetus hollandicus* was isolated from laboratory-scale nitrifying bioreactors and grown at a broad temperature range from 25–63 °C and belongs to the phylum *Chloroflexi*. To date, no further NOB have been identified in the genus *Chloroflexi*.[40]

10.8 Growth and Physiology of NOB

The media composition is defined as the metabolic flexibility of various NOB species. Preparing media by forming stock solutions after dilution is most suitable. The media component nitrite serves as the sole electron donor; however, to prevent toxic effects it is added as per the requirement of the NOB species. However, along with the media supplement, other factors affecting NOB growth are discussed in the sections below (Table 10.2).

10.8.1 Factors Influencing the Growth of NOB

The different factors affecting the growth of NOB are listed below:[41]

10.8.1.1 Volatile Fatty Acids

Acetic acid, *n*-caproic acid, formic acid, propionic acid, *n*-butyric, *n*-valeric, isobutyric, and isovaleric are inhibitory to nitrification.

10.8.1.2 Chlorine

NOB are resistant to monochloramines. Increased concentrations of oxidized nitrogen in combination will decrease the chlorine concentration in treated wastewater and it is used as an indicator for nitrification.

10.8.1.3 Light

NOB are recovered from dark conditions *i.e.*, photo-inhibition. Recovery of NOB is dose and light-wavelength-dependent. A porphyrin-like pigment (light-absorbing pigment) was found in NOB at the maximum absorption wavelength of 480 nm. These photoreceptors are seen in higher concentrations in AOB than NOB.

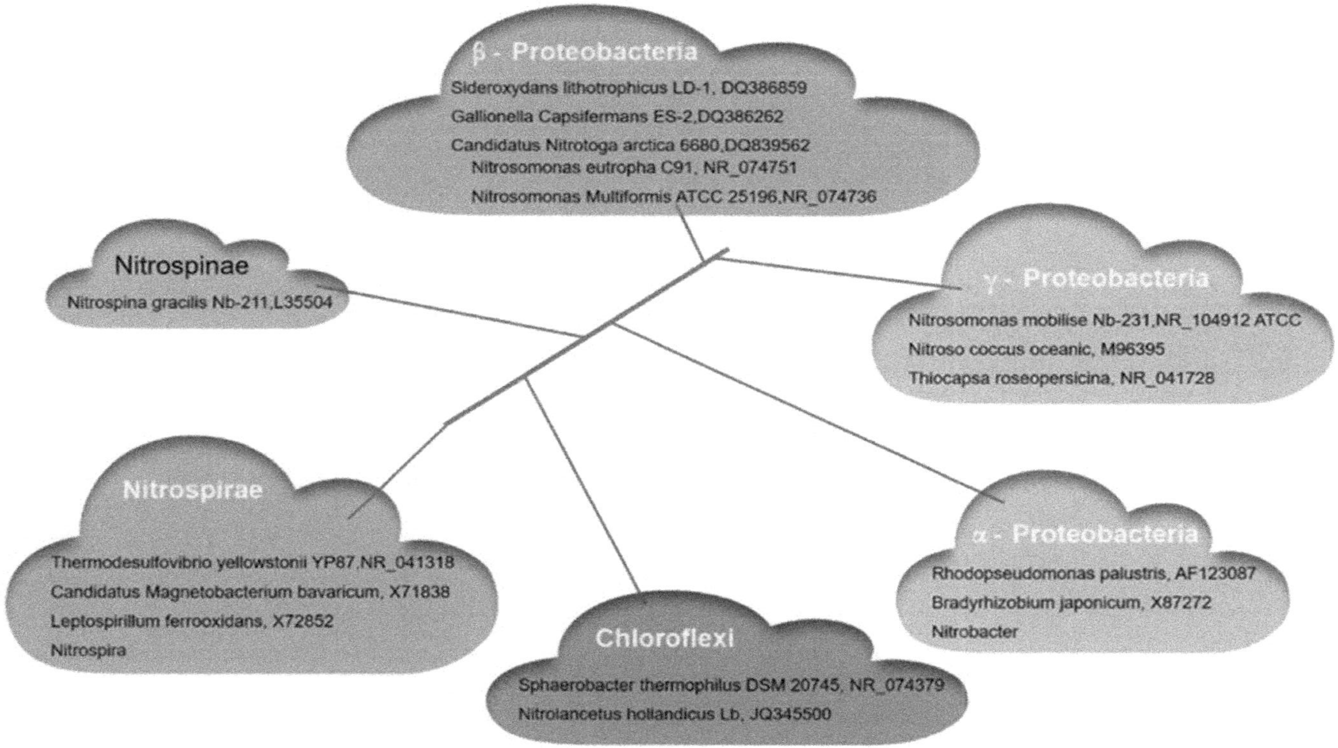

Figure 10.2 Phylogenetic analysis of NOB.

Table 10.1 Different lineages of *Nitrospira* species based on 16S rRNA analysis.

Sr. no.	Lineages	Source
1	Lineage I	Most abundant NOB group *Nitrospira* isolated from artificial ecosystems like WWTPs, and *Candidatus N. defluvii* from enrichment culture, one pure strain of ND1.[34,37]
2	Lineage II	*N. moscoviensis* from heating water, *N. japonica* from WWTP. Isolates from activated sludge, and habitants of soils and a freshwater system.[18,33,38]
3	Lineage III	*Nullarbor* cave system isolates and lacks cultivation history.[60]
4	Lineage IV	*N. marina* from deep-sea samples, sediments, and marine sponges.[61]
5	Lineage V	*N. bockiana* from heating water, few isolates from the soil. Recently identified lineages.[35]
6	Lineage VI	*N. calida* is a moderate thermophilic strain and a few strains from hot springs.[61]

Table 10.2 Optimal conditions for nitrification and nutrition required for NOB.

Condition	Optimum range for nitrification
Dissolved oxygen	>2 ppm
pH	7.5 to 8.0
Temperature	20–35 °C
Heavy metals	Nil
Toxic organics	Nil
Alkalinity	>100 ppm
Ca	0.5 mg L^{-1}
Cu	0.005–0.03 mg L^{-1}
Cr	Nil
Fe	7.0 mg L^{-1}
Mg	0.03 to 12.5 mg L^{-1}
Mb	0.001 to 1.0 mg L^{-1}
Ni	0.1 mg L^{-1}
P	310 mg L^{-1}
Zn	1.0 mg L^{-1}

10.8.1.4 Toluene

Higher concentrations of toluene reduce the nitrite oxidations after 45 days of incubation.

10.8.1.5 Trichloroethylene

Inhibits the nitrification of soil indigenous NOB populations.

10.8.1.6 Heavy Metals

Copper and nickel strictly inhibit the growth of NOB.

10.8.1.7 Alkalinity

Ammonia will inhibit the oxidation of nitrite. The molar ratio of carbonate alkalinity and ammonia will increase from 4.1 to 9.4, and the rate of ammonium is doubled; at the same time, the concentration of ammonia will be available freely in the culture medium, which is a selective inhibitor of Nitrobacter bacteria.

10.8.1.8 pH

Wastewater is an important parameter for nitrification activity. Nitrification occurs at higher pH values. NOB are very sensitive to low pH. The optimum pH range of the Nitrobacter genus is 7.5 to 8.0. For example, Soda Lake and soda soils with a pH of 9.5–11 led to the isolation of *Nitrobacter* sp. They form tiny colonies on the mineral salt agar at a pH of 10.

10.8.2 Physiological Investigations

The following physiological parameters were studied earlier by Sanders *et al.* (2006).

10.8.2.1 Short-term Nitrifying Activity

The kinetics of the aquatic system can be measured by using short-term nitrifying activity tests, but the value does not express the importance of *in situ* activity of NOB due to the aeration and additional nutritional substrate. However, it is important to measure the nitrite activity over a period that does not give the population time to increase significantly. To avoid this, the nitrite activity test should be more than 12 h. Under aerobic conditions, a decrease in nitrite concentration and an increase in the stoichiometry of nitrate will reflect the NOB activity. From this, the maximum activity was calculated as $g\,L^{-1}$ of sample per hour (Table 10.3).

10.8.2.2 Potential Nitrite-oxidizing Activity

To compare the potential of nitrification and the capacity of natural specimens, the substrate is incubated under various temperatures, which is favorable for nitrification, with the unknown sample as a control. Duplicate

Table 10.3 Fatty acid composition of *Nitrospira* species.

Fatty acid composition	Strains
16 : 10 11 methyl	*Nitrospira moscoviensis, Nitrospira calida, Candidatus Nitrospira bockiana*
16 : 1 *cis*-11 fatty acid	*Nitrospira moscoviensis, Nitrospira marina, Candidatus Nitrospira defluvii*

cultures are used to continuously monitor the substrate changes from nitrite to nitrate that are completely oxidized without any shaking to measure the potential nitrite-oxidizing activity. Organic matter and micro-niches will favour the denitrification process (Sanders *et al.*, 2006). So, the test vessels must be shaken for continuous oxygen supply and to analyse the essential formation of nitrate and disappearance of nitrite.

10.8.2.3 Growth Parameters

During the growth of AOB and NOB in batch culture, the concentration of product increases exponentially, so the growth is balanced under this assumption. The protein content determination is possible only when the bacteria are grown in mixotrophic culture conditions.[42] The BCA protein content determination method is more suitable than the Bradford method and is high sensitivity.

Bacterial cell quantification is very difficult when bacterial cells are aggregated to form microcolonies. In addition, the *Nitrospira* cells are very small compared to other NOB genera and difficult to detect by using a counting chamber with a 40× objective under a microscope. This can be achieved by using a thin counting chamber with a 100× objective under a microscope.

Growth experiments of NOB can be achieved by the determination of temperature, which requires optimal growth, pH, nitrite conditions, *etc.* To start the procedure, 40 mL of mineral medium containing 0.75 mM nitrate is transferred into 2 mL aliquots containing active pre-culture, and the tubes are incubated without stirring with varying temperatures of 4 °C and 37 °C. Nitrite oxidation is optimal at 32 °C for 7 days. No nitrite oxidation activity occurs at 4 °C or 37 °C, whereas less nitrite oxidation activity occurs at 10 °C.

10.8.2.4 Purity Test

The purity test was done as per earlier studies.[43]

- Liquid Medium: 5 mL of the liquid medium is transferred to small aliquots and one drop of the culture is added to check the turbidity for up to 2 weeks.
  - Media Composition: peptone, yeast extract, meat extract, NaCl, pH 7.3 to 7.4.
- Solid Medium: 15 grams of agar per litre of sample to check the turbidity.
  - Media Composition: meat extract, casamino acid, yeast extract, NaCl, potassium dihydrogen phosphate, and pH should be 7.3 to 7.4.
- Oligotrophic Purity Tests: the use of a mixotrophic medium to check the presence of heterotrophic oligotrophs without any nitrite. Slight turbidity and absence of colonies after several weeks of incubation indicate the growth of heterotrophic NOB. Colony formation and heavy turbidity indicate contamination of other heterotrophs.

10.8.2.5 Nutritional Requirements

Lithoautotrophic nitrite oxidizers are isolated from the mineral nitrite media containing free organic contaminants. If organic contaminants are present, the overgrowth of heterotrophic bacteria is seen in the culture plates. The various media are developed by different scientists as follows.

10.8.2.6 Mineral Salt Medium

The growth of lithoautotrophic bacteria was developed by Bock *et al.*[27] The medium contains $CaCo_3$, NaCl, $MgSO_4$, KH_2PO_4, and trace elements including $MnSO_4$, 0.01 M HCl, H_3BO_3, $ZnSO_4$, $FeSO_4$, $CuSO_4$, $(NH_4)_6Mo_7O_{24}$, $FeSO_4$, $CuSO_4$, pH 8.4 to 8.6. Metal or plastic caps or cotton stoppers are used and the medium is prepared 2 days before so that the pH of the medium can be maintained and it can be saturated with carbon dioxide.[27]

- The Marine Medium: this medium contains seawater, $CaCl_2$, $MgSO_4$, $FeSO_4$, KH_2PO_4, and marine trace elements. The pH of the medium is 6.5 to 7.0.[9]
- Mixotrophic Medium: medium containing yeast extract, peptone, sodium pyruvate, and $NaNO_2$; the pH of the medium is 7.4 to 7.6.[27]
- Heterotrophic Medium: the medium contains yeast extract, peptone, sodium pyruvate, and mineral salt medium constituents. The pH of the medium should be maintained at 7.4.[27]
- Acidic Medium: media for the isolation of the *Nitrobacter* strain growing at pH 5.5. This medium contains KH_2PO_4, K_2PO_4, $FeSO_4$, Na_2 EDTA, $CaCl_2$, $MgSO_4$, and the trace elements of $NaMoO_4$, $MnCl_2$, $CoCl_2$, $ZnSO_4$ and $CuSO_4$, and the pH is 5.0.[44]

10.9 Isolation Techniques of NOB

10.9.1 Cultivation of NOB

For environmental samples, 1 mL or 1 g of the sample is used for 150 mL of the medium in 300 mL flasks. Incubation of the culture must be carried out under dark conditions without stirring and the temperature varies depending on the bacterial strain.[45] The consumption of nitrite by bacteria can be checked by using a modified Griess-Ilosvay spot test.

10.9.2 Modified Griess-Ilosvay Spot Test

Schmidt and Belsar performed a test by placing a drop of the liquid culture on filter paper containing a reagent that produces a pink spot when the culture is not active.[46] If the culture is active, it does not produce a pink colour. The nitrite concentration is refilled by using a 2.5 M stock solution of nitrite.

Continuous feeding of nitrite may result in increasing concentration, which may lead to inhibition of the culture.

10.9.3 Reagents of the Modified Griess-Ilosvay Spot Test

It contains 2 reagents that should be stored at 4 °C in dark conditions.

Solution-I (diazotizing reagent) contains sulfanilamide, and orthophosphoric acid.

Solution-II (coupling reagent) is *N*-(1-naphthyl)-ethylenediamine-dihydrochloride.

10.9.4 Plating Technique

NOB are sensitive to oxygen; therefore, growth is not visible on the surface of the solid medium. However, two species of the *Nitrobacter* genus are easily cultivated and seen in the colonies on the surface of the agar.[19,27] The colonies are small, shiny, and brownish with entire growth on a mixotrophic medium observed after several days of incubation in the laboratory. Plates should maintain moisture content to prevent drying.

10.9.5 Dilution of the Sample

A mineral medium containing nitrite is used to separate NOB from a heterotrophic environment.[45] Usually, low nitrate content is required for most of the NOB to grow. Samples are diluted serially by using an enrichment medium and tubes are incubated for several days to months and the consumption of nitrite is assayed periodically for successful isolation.

10.9.6 Storage of NOB

Liquid cultures can be stored at the recommended temperature of 17 °C in a dark environment and cells can be reactivated after several years.[46,47] The genus *Nitrospira* is extremely sensitive and exceedingly difficult to store; however, the beginning concentration of sodium nitrite can be lowered for successful storage, especially in marine strains, *i.e.*, 0.15 mM concentration. Liquid nitrogen freezing is the most effective method for the maintenance of the stock cultures by inoculating the cultures in cryoprotecting buffer Hatefi. The components of the Hatefi buffer are Tris, sucrose, and histidine, and the pH is 7.5. In addition, the storage of *Nitrobacter* by using a mixotrophic medium with an increased amount of organic matter in the medium is multiplied by a factor of 10. Glycerol stock is used instead of using pyruvate for the storage of *Nitrobacter* sp. as the glycerol keeps the pH stable for a long duration.

10.9.7 Most Probable Number (MPN)

It is a conventional method used for the enumeration of bacterial counts, but maximal yields have not been resolved by using the MPN technique. The MPN method requires longer incubation, approximately 103–113 days for NOB identification. pH and nitrite concentration influence the MPN enumeration results. For example, NOB in soil requires low nitrite concentration for maximal cell count yield.[41]

Nitrifying bacteria are enumerated by using the MPN technique with the modified media of Alexander and Clark. The chemical composition of Alexander and Clark media is $(NH_4)_2SO_4$ (0.5 $g L^{-1}$), K_2HPO_4 (1.0 $g L^{-1}$), $FeSO_4 \cdot 7H_2O$ (0.03 $g L^{-1}$), NaCl (0.3 $g L^{-1}$), $MgSO_4 \cdot 7H_2O$ (0.3 $g L^{-1}$), $CaCO_3$ (7.5 $g L^{-1}$), pH 8.2. Samples like soils, manure, sewage effluent stock solutions, *etc.* are diluted by tenfold series in normal physiological saline. Test tubes are visually observed at 5–8 weak intervals at 30 °C, and positive samples are indicated by a physical change in the colour *i.e.*, production of acid is indicated by an indicator incorporated in the medium. To check NOB present in the sample, the α-naphthylamine, sulfanilic acid, and sodium acetate method can be used. Denitrifying bacteria are enumerated by using the GC-MPN method. The headspace of the test tube contains gas produced by the activity of bacteria, the gas is replaced with nitrogen and incubated at 30 °C and the N_2O gas accumulation is measured by using GC (Gas Chromatography). The equipment is equipped with an electron capture detector that detects automatically and does not require any chemicals for any further identification.

10.10 Alternative Metabolisms of NOB

Previously only the nitrite oxidation activity of NOB was known. Although, a few *Nitrobacter* and *Nitrospira* members were found to benefit from the presence of simple organic compounds along with nitrite and CO_2.[28,32,48] However, some *Nitrobacter* species can grow in heterotrophic mode only.[49] In the natural environment, the occurrence of NOB is usually linked to the nitrification process. The genome analysis has changed this scenario and revealed that the true metabolic activity of NOB occurs in various environmental sources. It is known that *N. moscoviensis* codes for hydrogenase and other accessory proteins that play a role in the maturation and transcriptional regulation of hydrogenases.[50] Considering this ability, *N. moscoviensis* was grown by providing H_2 as a sole source of energy and electron donor, O_2 as an electron acceptor, and CO_2 as a single carbon source. This study revealed that *N. moscoviensis* was able to grow aerobically by the respiration of H_2. In this way, the first NOB exterior to the nitrogen cycle was reported.[50] Although hydrogen was confirmed to be consumed at nano-molar concentration, the lowest concentration of H_2 utilized by *N. moscoviensis* still remains unclear. It was also confirmed that *N. moscoviensis* can consume H_2 anaerobically along with nitrite, which is an electron acceptor; however, NOB growth was unnoticed.[33]

10.11 Temperature Effect on NOB

Substrate concentration and temperature have a significant role in the assortment of NOB in natural ecosystems or habitats. For the case study of a municipal WWTP, a significant alteration in the balance of *Nitrosomonas* and *Nitrosospira* microorganisms due to variations in seasonal temperature occurs. Detailed soundings of nitrification in cold ecosystems are rare, even though it is well-recognized in soils at near-freezing temperatures. The strain of *Nitrosomonas cryotolerans*, isolated from Alaskan waters, nitrifies below −5 °C, even though the ideal range of temperature is between 22 °C and 30 °C. With the geographical area and climate, the optimum temperature range of NOB and ammonia varies (*e.g.*, 35 °C in tropical soil and for moderate 20 °C). It is assumed that differences in temperature flexibility are because of unstable populations.[41]

The annual temperature range in the WWTP of Dradenau/Köhlbrandhöft Hamburg typically ranges between 12.2 °C and 21.2 °C. Nitrification problems arise during the winter season when the operational temperature decreases below 10 °C. The nitrification rate reduces by 30% at 10 °C and at 5 °C it is reduced from 8% to 31%. Whereas at 17 °C the process of denitrification has already been reduced. Culture and molecular-based studies of nitrifying microbes isolated from WWTPs and bioreactors specified that NOB belong to the *Nitrospira* genus, and they were widespread. Previously, *Nitrobacter* was considered the chief NOB, and later *Nitrobacter* and *Nitrospira* were isolated in a laboratory at 28 °C. The research conducted on the nitrogen cycle in permafrost pretentious soils in Siberia led to the supposition that an innovative beta proteobacterium is the chief organism accountable for the oxidation of nitrite in this ecosystem. The detection of the low-temperature adapted '*Candidatus Nitrotoga Arctica*' specified that the ability of bacteria to oxidize nitrite is more extensive than commonly presumed. Unexpectedly, the sequence of 16S rRNA of this organism was closely associated with a clone AB247475 that existed in domestic sewage with a distinct affinity for polyphosphate accumulation. These conclusions provoked an investigation into whether bacteria like *Nitrotoga* may also live in moderate weather WWTPs of Europe. In this research, physiological, phylogenetic, and morphological parameters indicate that supplementary to '*Candidatus Nitrospira defluvii*' and unique *Nitrobacter* such as clone Drad-1, a novel *Nitrotoga* bacterium like clone Ham-1 also existed in the Hamburg WWTP. The above results give intuition that the NOBs are adapted to cold temperatures.[40]

10.12 A Newer Technique for the Identification and Cultivation of NOB

10.12.1 Amperometric Enzyme-linked Immunoassay

This technique is used for the enumeration of Nitrobacter species. It is an immunological reaction in which a glassy carbon electrode is used. The limit of detection (LOD) of *Nitrobacter* is approximately 3×10^{-6} bacterial cells per mL.[11]

10.12.2 Fatty Acid Analysis

This is a useful technology for the rapid identification and differentiation of NOB. At present five known genera of NOB are available, and these can be differentiated by using various biomarkers of lipid components present on the cell wall of the bacteria. The genus *Nitrobacter* contains *cis*-11-octadecenoic acid (18 : 1 *cis*-11), which accounts for 70% of the total fatty acids in the cell wall, and the remaining 30% contains *cis*-11–12 methyloctadecanoic acid (19 : 0 cyclo 11–12). The sum of both fatty acids (*cis*-11-octadecenoic acid and *cis*-11–12 methyloctadecanoic acid) present on the cell wall of the *Nitrobacter* is 80–90%. Based on the growth phase and growth conditions, the ratio between lipid components varies. The *Nitrobacter* species do not have hydroxyl fatty acids.[34,51] Gammaproteobacteria of the *Nitrococcus* genus have a very low amount of 18 : 11 *cis*-11 compared to *Nitrobacter* and a high amount of hydroxyl fatty acids, whereas the genus *Nitrospira* contains *cis*-7 and *cis*-11 isomers of hexadecenoic acid and 11-methylhexadecanoic acid (16 : 10 11 methyl). 16 : 10 11 methyl are branched-chain lipids, and they are unique in the *Nitrospira* genus. However, *cis*-9 hexadecenoic acid is a common membrane lipid component of most of the bacteria but it is very rarely seen in the *Nitrospira* strains. Combinations and concentrations of the above three lipid biomarkers can be used for the chemotaxonomic identification of *Nitrospira* at the species level as follows.

10.12.3 Procedure for Fatty Acid Analysis

The centrifugation of dense brownish floccus turbid bacterial cultures at 15 000 × *g* for 30 min is performed to collect bacterial cell pellets. The pellet-containing cells are washed with 0.9% NaCl and stored at −20 °C. Approximately 20–50 mg of wet pellet is transferred to a Teflon-lined screw-capped tube and 15% of 1 mL of NaOH is added in a 1 : 1 ratio of methanol/water and the tubes are incubated at 100 °C for 30 min for cell lysis and saponification of fatty acids. From the extracted fatty acid methyl esters (FAME), 2 mL of 6.0 N HCl/methanol (1.2 : 1 v/v) ratios are added and the tubes are incubated at 80 °C for 10 min for fatty acid methylation and 1.25 mL of 1 : 1 (v/v) ratio of hexane/methyl *tert*-butyl ether is added followed by end-over-end mixing for 10 min in a tube rotator. The lower aqueous phase is discarded after the phase separation; 3 mL of 1.2% NaOH is added to the remaining organic phase for washing by an end-over-end rotator for 5 min. Finally, the phase separation is supported by centrifugation at 1000 × *g* for 3 min, and the upper clear liquid that contains the organic phase is transferred to the GC (gas chromatograph) for analysis.

10.12.4 Electron Microscopy

The liquid culture is centrifuged at 15 000 × *g* and the cell pellet is collected, followed by addition of 0.9% NaCl for washing, and prefixing of the cells with 2.5% glutaraldehyde for 30 min at RT. The cells are washed three times by using cacodylic acid and 2% noble agar is used for embedding, then 1% (w/v) of osmium tetroxide (buffered with cacodylic acid) is used for fixation at 4 °C overnight followed by four times washing with cacodylic acid. Different concentrations of acetone (15%, 30%, 50%, 70%, 90%, and 100%) are used for dehydration at 4 °C for 10 min each. Different ratios of spur and acetones are used for embedding, *i.e.*, 1 : 3 for 2 h, 1 : 1 for 2 h, and 3 : 1 for 17 h, followed by the addition of pure spur for 2.5 h and incubating the tubes at 6 h in freshly prepared spur. The samples are covered with spur and polymerized for 16 h at 70 °C to appear in silicone forms. The dried samples are sectioned with a diamond knife and kept on copper grids for staining using 2% of uranyl acetate in distilled water and 2% lead acetate for 10 min. These samples are observed under TEM (transmission electron microscope).[41]

10.12.5 Molecular Assays

PCR (Polymerase Chain Reaction) is applied to detect and quantify the nitrifying bacteria by using species-specific primers for the genus. The labelling of ^{14}C radio-tracking is used to measure the nitrification potential.[52] Two primers for the *Nitrobacter* population were taken from the 16S rRNA segment at a size of 397 gene fragments for amplification and to detect a lower threshold in PCR (10^2 cells per g of soil).[53]

10.12.6 FISH (Fluorescent *In Situ* Hybridization) Assay

Highly specific oligonucleotide probes are used and targeted against the 16s rRNA of single nucleotide polymorphisms (SNPs) that are closely related to nitrifiers.[54] Single-cell identification and enumeration are observed by using epifluorescence microscopy or cells can be differentiated by using flow cytometry.

10.12.7 Catalyzed Reporter Deposition-FISH (CARD-FISH)

The application of a hybridization signal will give more sensitivity. The hybridization signal is based on the activity of Horseradish peroxidase (HPR) linked to an oligonucleotide probe. It will result in the deposition of the fluorescence-tagged tyramide molecules with HRP conjugation. Activated target cells bind to tyramide molecules permanently so the fluorescent signal is strong and stable.

10.12.8 Monoclonal Antibodies

Fluorescent monoclonal antibodies can be used for the rapid quantification and identification of specific NOB *in situ* from a mixed bacterial culture.[55] Genus-specific monoclonal antibodies are used to recognize nitrite oxidoreductase (NOR). NOR is common in *Nitrobacter* species, and monoclonal antibodies are targeted and designed to recognize the epitope of the NOR α-subunit. For example, the monoclonal antibody Hyb 153-1 is designed for the identification of *Nitrobacter hamburgensis*, Hyb 153-2 for *Nitrobacter vulgaris*, Hyb 153-3 *Nitrobacter winogradskyi*, *etc.*

10.13 Methods for Measurement of Nitrification Rates

There are several methods for the measurement of nitrification rates.[56,57]

10.13.1 ^{15}N Tracer Methods

An effective method for quantification and measurement of nitrification rates by using a ^{14}N tracer. In this method, a small amount of $^{15}NH_4^+$ or $^{15}NO_2^-$ is added to the sample and incubated, and the extracted labelled product $^{15}NO_2^-$ or NO_3^- is measured by spectrophotometry. The main advantage and sensitivity of this method are to compare the samples with inhibitor methods like duration of incubation, and *in situ* conditions (Nutrients, Sunlight, *etc.*). However, inhibitor-based assays give more advantages than other methods.

10.13.2 Inhibitor-based Nitrification Assay (IBNA)

IBNA depends on the ability of several compounds to interact with active sites of the nitrite oxidoreductase (NXO). Potential inhibitors are added to NXO and incubations are carried out in dark conditions with replicates. Nitrification rates are measured by the change of N concentration and the presence or absence of CO_2 fixation measurement with inhibitors are commonly used approaches for this assay. Conversion of CO_2 and N_2 oxidation rates are used to estimate the nitrification rate. Scintillation counters are more common and easier to use than other instruments required for stable isotope analysis and inorganic N determinations.

10.13.3 Immuno-blot Immunoassays

Three monoclonal antibodies are used in immuno-blot assays that recognize the NXR (Nitrite oxidoreductase). The monoclonal antibodies are directed against the Nitrite oxidizing system (NOS) of *Nitrococcus*, *Nitrospira*, and *Nitrospira*.[58] It is similar to the monoclonal antibodies test.

10.14 Chemotaxonomy

NOB gain energy from the oxidation of nitrite to nitrate, which is required for growth. The important key enzyme required for nitrite oxidation is nitrite oxidoreductase (NXR). The enzyme NXR is responsible for the transfer of 2 electron peroxidized NO_2^- to the electron transport chain. NXR is an

Intracytoplasmic Membrane (ICM) catabolic enzyme, and *Nitrobacter* and *Nitrococcus* have ICM that requires a nitrate and nitrite antiport transport system. The genera *Nitrospira*, *Nitrospina*, and *Nitrotoga* lack ICM and are more sensitive to NXR. NXR is present in the periplasmic space of the cytoplasmic membrane so it does not require a nitrite transport system.[56] NXR has three membrane structural subunits, namely:

10.14.1 Alpha Subunit

Substrate binding site (NxrA) with a molybdopterin co-factor.

10.14.2 Beta-subunit

Codes for the NxrB gene, which transfers an electron from NxrA to the gamma subunit of the membrane integral.

10.14.3 Gamma Subunit

It is a holoenzyme and channels and is anchored in the cell membrane. It transfers electrons to the electron transport chain. However, the exact components and compositions of the structural complex NXR of NOB are still awaiting experimental confirmation. NOB has the enzyme cyanase which converts cyanate [OCN^-] to NH_3 and CO_2. Within bacterial cells, the formation of cyanate is mainly due to the spontaneous degradation of carbamoyl phosphate. Pools of environmental cyanate may lead to the physiochemical decomposition of urea and cyanide. Co-culture of the AOB (*Nitrosomonas nitrosa*) is performed with cyanase positive and NOB (*N. moscoviensis*) which convert nitrified cyanate to nitrate by reciprocal feedings. However, further research is required to address the ecological significance.

10.15 Conclusion

The prevention of nitrite accumulation in water bodies is essential to avoid the occurrence of health hazards among biotic populations and environmental changes like eutrophication, the greenhouse effect, *etc.* Several physical, chemical, and biological methods are currently in practice to remove heavy metals and toxic organics like nitrite present in industrial wastewater. The chemical method for nitrite degradation has little success, and hence it has limited application for the successful removal of nitrite present in industrial wastewater. Hence, research has focused on the use of microbes to biotransform nitrite present in industrial wastewater. This biological method is an eco-friendly and low-cost approach to resolving the environmental issues of nitrite accumulation. In this process, there is the oxidation of ammonia to nitrite and then nitrate. The balance between AOB and NOB plays an important role in the optimization of the nitrifying community. Further research can be made possible to study the physiological activity and molecular mechanisms of NOB. The computational study combined with wet-lab experiments can be very useful to understand the effective nitrification process and the biogeochemical cycle of nitrification. Furthermore, the main hurdle in this path is the difficulty in the cultivation and enumeration of NOB. Hence, there is a necessity to update the technical skills and experimental protocols for the successful identification of NOB and study their role in the nitrification process.

References

1. R. Gurav, S. K. Bhatia, T.-R. Choi, Y.-K. Choi, H. J. Kim, H.-S. Song, S. M. Lee, S. Lee Park, H. S. Lee, J. Koh, J.-M. Jeon, J.-J. Yoon and Y.-H. Yang, *Chemosphere*, 2021, **264**, 128539.
2. S. S. Suryawanshi, P. P. Kamble, R. Gurav, Y.-H. Yang and J. P. Jadhav, *Biomass Convers. Biorefin.*, 2023, **13**, 5353–5366.
3. G. Vyavahare, P. Jadhav, J. Jadhav, R. Patil, C. Aware, D. Patil, A. Gophane, Y.-H. Yang and R. Gurav, *J. Cleaner Prod.*, 2019, **207**, 296–305.
4. G. Vyavahare, R. Gurav, R. Patil, S. Sutar, P. Jadhav, D. Patil, Y.-H. Yang, J. Tang, C. Chavan, S. Kale and J. Jadhav, *Environ. Geochem. Health*, 2021, **43**, 2913–2926.
5. Y.-K. Choi, T.-R. Choi, R. Gurav, S. K. Bhatia, Y.-L. Park, H. J. Kim, E. Kan and Y.-H. Yang, *Sci. Total Environ.*, 2020, **710**, 136282.
6. R. Gurav, S. K. Bhatia, T.-R. Choi, Y.-K. Choi, H. J. Kim, H.-S. Song, S. L. Park, H. S. Lee, S. M. Lee, K.-Y. Choi and Y.-H. Yang, *Sci. Total Environ.*, 2021, **781**, 146636.
7. J. E. Kim, S. K. Bhatia, H. J. Song, E. Yoo, H. J. Jeon, J.-Y. Yoon, Y. Yang, R. Gurav, Y.-H. Yang, H. J. Kim and Y.-K. Choi, *Bioresour. Technol.*, 2020, **306**, 123092.
8. R. Gurav, S. K. Bhatia, T.-R. Choi, H. J. Kim, Y.-K. Choi, H.-J. Lee, S. Ham, J. Y. Cho, S. H. Kim, S. H. Lee, J. Yun and Y.-H. Yang, *Chemosphere*, 2022, **296**, 134034.
9. S. W. Watson and J. B. Waterbury, *Arch. Mikrobiol.*, 1971, **77**, 203–230.
10. G. J. Both and H. J. Laanbroek, *FEMS Microbiol. Lett.*, 1991, **85**, 335–344.
11. S. Bartosch, I. Wolgast, E. Spieck and E. Bock, *Appl. Environ. Microbiol.*, 1999, **65**, 4126–4133.
12. M. Alawi, A. Lipski, T. Sanders, E. M. Pfeiffer and E. Spieck, *ISME J.*, 2007, **1**, 256–264.
13. M. Alawi, S. Off, M. Kaya and E. Spieck, *Environ. Microbiol. Rep.*, 2009, **1**, 184–190.
14. E. V. Lebedeva, M. Alawi, C. Fiencke, B. Namsaraev, E. Bock and E. Spieck, *FEMS Microbiol. Ecol.*, 2005, **54**, 297–306.
15. S. Winogradsky, *Contributions a la morphologie des organismes de la nitrification*, 1892.
16. N. Gruber, *Nitrogen Mar. Environ.*, 2008, **2**, 1–50.

17. A. G. Castellani and C. F. Niven, Jr., *Appl. Microbiol.*, 1955, **3**, 154–159.
18. N. Ushiki, H. Fujitani, Y. Aoi and S. Tsuneda, *Microbes Environ.*, 2013, **28**, 346–353.
19. D. Y. Sorokin, G. Muyzer, T. Brinkhoff, J. G. Kuenen and M. S. Jetten, *Arch. Microbiol.*, 1998, **170**, 345–352.
20. T. H. Blackburn and K. Henriksen, *Limnol. Oceanogr.*, 1983, **28**, 477–493.
21. K. Inomura, J. Bragg, L. Riemann and M. J. Follows, *PLoS One*, 2018, **13**, e0208282.
22. A. Schramm, D. De Beer, A. Gieseke and R. Amann, *Environ. Microbiol.*, 2000, **2**, 680–686.
23. S. Pajares and R. Ramos, *Front. Mar. Sci.*, 2019, **6**, 739.
24. M. A. Rinaldi, C. A. Ferraz and N. S. Scrutton, *Nat. Prod. Rep.*, 2022, **39**, 90–118.
25. J. F. Ortiz-Medina, A. M. Grunden, M. R. Hyman and D. F. Call, *ACS Sustainable Chem. Eng.*, 2019, **7**, 3511–3519.
26. D. Imadud, *J. Saudi Soc. Agric. Sci.*, 2021, **20**, 459–466.
27. E. Bock, H. Sundermeyer-Klinger and E. Stackebrandt, *Arch. Microbiol.*, 1983, **136**, 281–284.
28. E. Bock, H.-P. Koops, U. C. Möller and M. Rudert, *Arch. Microbiol.*, 1990, **153**, 105–110.
29. B. Vanparys, E. Spieck, K. Heylen, L. Wittebolle, J. Geets, N. Boon and P. De Vos, *Syst. Appl. Microbiol.*, 2007, **30**, 297–308.
30. E. Spieck, S. Keuter, T. Wenzel, E. Bock and W. Ludwig, *Syst. Appl. Microbiol.*, 2014, **37**, 170–176.
31. J. M. Beman, J. Leilei Shih and B. N. Popp, *ISME J.*, 2013, **7**, 2192–2205.
32. S. W. Watson, E. Bock, F. W. Valois, J. B. Waterbury and U. Schlosser, *Arch. Microbiol.*, 1986, **144**, 1–7.
33. S. Ehrich, D. Behrens, E. Lebedeva, W. Ludwig and E. Bock, *Arch. Microbiol.*, 1995, **164**, 16–23.
34. E. Spieck, C. Hartwig, I. McCormack, F. Maixner, M. Wagner, A. Lipski and H. Daims, *Environ. Microbiol.*, 2006, **8**, 405–415.
35. E. V. Lebedeva, M. Alawi, F. Maixner, P. G. Jozsa, H. Daims and E. Spieck, *Int. J. Syst. Evol. Microbiol.*, 2008, **58**, 242–250.
36. S. Keuter, M. Kruse, A. Lipski and E. Spieck, *Environ. Microbiol.*, 2011, **13**, 2536–2547.
37. H. Fujitani, N. Ushiki, S. Tsuneda and Y. Aoi, *Environ. Microbiol.*, 2014, **16**, 3030–3040.
38. R. Seviour and P. H. Nielsen, *Microbial ecology of activated sludge*, IWA publishing, 2010.
39. J. I. Schwarz, W. Eckert and R. Conrad, *Syst. Appl. Microbiol.*, 2007, **30**, 239–254.
40. D. Y. Sorokin, S. Lücker, D. Vejmelkova, N. A. Kostrikina, R. Kleerebezem, W. I. C. Rijpstra, J. S. S. Damsté, D. Le Paslier, G. Muyzer, M. Wagner, M. C. M. van Loosdrecht and H. Daims, *ISME J.*, 2012, **6**, 2245–2256.

41. A. Abeliovich, in *The Prokaryotes: Volume 5: Proteobacteria: Alpha and Beta Subclasses*, ed. M. Dworkin, S. Falkow, E. Rosenberg, K.-H. Schleifer and E. Stackebrandt, Springer New York, New York, NY, 2006, pp. 861–872.
42. J. I. Prosser, *Adv. Microb. Physiol.*, 1989, **30**, 125–181.
43. W. Steinmüller and E. Bock, *Arch. Microbiol.*, 1976, **108**, 299–304.
44. T. R. Hankinson and E. L. Schmidt, *Appl. Environ. Microbiol.*, 1988, **54**, 1536–1540.
45. E. Spieck and A. Lipski, *Methods Enzymol.*, 2011, **486**, 109–130.
46. E. L. Schmidt and L. W. Belser, Autotrophic nitrifying bacteria, in *Methods of soil analysis*, ed. R. W. Weaver *et al.*, SSSA Book Ser. 5. Part 2, SSSA, Madison, WI, 1994, pp. 159–177.
47. A. Freitag, M. Rudert and E. Bock, *FEMS Microbiol. Lett.*, 1987, **48**, 105–109.
48. H. Daims, J. L. Nielsen, P. H. Nielsen, K. H. Schleifer and M. Wagner, *Appl. Environ. Microbiol.*, 2001, **67**, 5273–5284.
49. E. Bock, *Arch. Microbiol.*, 1976, **108**, 305–312.
50. H. Koch, A. Galushko, M. Albertsen, A. Schintlmeister, C. Gruber-Dorninger, S. Lücker, E. Pelletier, D. Le Paslier, E. Spieck, A. Richter, P. H. Nielsen, M. Wagner and H. Daims, *Science*, 2014, **345**, 1052–1054.
51. A. Lipski, E. Spieck, A. Makolla and K. Altendorf, *Syst. Appl. Microbiol.*, 2001, **24**, 377–384.
52. T. Berthea, J. Garnier and F. Petit, *C. R. Acad. Sci. III*, 1999, **322**, 517–526.
53. V. Degrange and R. Bardin, *Appl. Environ. Microbiol.*, 1995, **61**, 2093–2098.
54. C. Gruber-Dorninger, M. Pester, K. Kitzinger, D. F. Savio, A. Loy, T. Rattei, M. Wagner and H. Daims, *ISME J.*, 2015, **9**, 643–655.
55. N. Noda, H. Ikuta, Y. Ebie, A. Hirata, S. Tsuneda, M. Matsumura, T. Sumino and Y. Inamori, *Water Sci. Technol.*, 2000, **41**, 301–308.
56. H. Daims, S. Lücker and M. Wagner, *Trends Microbiol.*, 2016, **24**, 699–712.
57. B. B. Ward, *Methods Enzymol.*, 2011, **486**, 307–323.
58. A. P. Rees, E. M. S. Woodward, C. Robinson, D. G. Cummings, G. A. Tarran and I. Joint, *Deep Sea Res., Part II*, 2002, **49**, 2905–2927.
59. N. Lehnert, B. W. Musselman and L. C. Seefeldt, *Chem. Soc. Rev.*, 2021, **50**, 3640–3646.
60. S. Off, M. Alawi and E. Spieck, *Appl. Environ. Microbiol.*, 2010, **76**, 4640–4646.
61. E. V. Lebedeva, S. Off, S. Zumbrägel, M. Kruse, A. Shagzhina, S. Lücker, F. Maixner, A. Lipski, H. Daims and E. Spieck, *FEMS Microbiol. Ecol.*, 2011, **75**, 195–204.

CHAPTER 11

Quantitative Methodologies for Determining the Amount and Structure of AOB at the Transcriptional Level in Wastewater Treatment Plants

SUBHAM PREETAM,[a] RAJESWARI RATH,[b] ISHANEE MAZUMDER,[c] SOFIA KHAN,[d] CHANDRASHISH ROY,[c] AMANAT ALI[e] AND SUMIRA MALIK*[f]

[a] Institute of Advanced Materials, IAAM, Gammalkilsvägen 18, Ulrika 59053, Sweden; [b] Centre for Biotechnology, Siksha O Anusandhan (SOA-DU), Bhubaneswar, Odisha, India; [c] KIIT School of Biotechnology, Kalinga; Institute of Industrial Technology (KIIT-DU), Bhubaneswar, Odisha, India; [d] Department of Chemistry Mohammad Ali Jauhar University, Rampur, India; [e] Department of New Materials and Nanotechnology, National University of Science and Technology MISiS, Moscow, Russia; [f] Amity Institute of Biotechnology, Amity University Jharkhand, Ranchi, Jharkhand, 834001, India
*Email: smalik@rnc.amity.edu

11.1 Introduction

Wastewater treatment involves a major biological nitrogen removal process, which has a significant effect on nitrogen compounds in the marine environment and wastewater. The primary purpose of aerobic activated sludge systems is to extract organic waste. Nitrification is a method where ammonia

Ammonia Oxidizing Bacteria: Applications in Industrial Wastewater Treatment
Edited by Maulin P. Shah

Published by the Royal Society of Chemistry, www.rsc.org

is oxidized to nitrate *via* nitrite, which helps in the biological removal of nitrogen in wastewater treatment systems. Unlike municipal WWTPs, some industrial WWTPs add nitrogen (mainly ammonia) to the influent to balance the high organic carbon waste stream and provide nitrogen for heterotrophic growth. Ammonia oxidizing bacteria (AOB) are vital in converting ammonia to nitrate in WWTPs. In this context, an understanding of the ecology and microbiology of AOB in wastewater treatment is needed to enhance treatment control.[1,2] The AOB require genomic methods that are not easy to isolate from the environment. Temperature is a core factor in biological nitrogen removal.

The abundance of AOB activity decreases and the nitrifying efficiency reaches a limit due to low temperature, which inhibits the AOB activity. AOB play a significant role in the conversion, giving soluble microbial products and nitrous oxides, and producing nitrate by using nitrite oxidizing bacteria (NOB). Ammonia monooxygenase (AMO) is a crucial enzyme in the biochemistry of AOA and AOB. Methane (CH_4), carbon monoxide (CO), and other aliphatic and aromatic molecules may be oxidized and function as competing inhibitors of bacterial AMO. The two most persistent factors of ammonia oxidizer variety and abundance are ecological pH and ammonia level.

Some BNR processes include single high ammonia removal over nitrite (SHARON) and anaerobic ammonium oxidation (ANAMMOX). The only physiology known for members of the Thaumarchaeota is chemolithoautotrophic growth on NH_3. It was recently discovered that Thaumarchaeota that show putative amo genes were considerably abundant in various wastewater treatment plant sludges.[3–5]

11.2 Ecological Niche and Physiology of Ammonium and Nitrogen Oxidizers

The evolutionary relationships of the chemolithoautotrophs of the nitrogen cycle are important to wastewater treatment.[6,7] To assess nitrifying population structures in activated sludge, several combinations of 16S rRNA-targeted oligonucleotide probes were utilized.

11.2.1 Ecological Niche of Ammonium Oxidizers

Ammonia-oxidizing bacteria (AOB) have been recognized for over a century to promote the aerobic oxidation of ammonia (NH_3) to nitrite, the first stage in nitrification (NO_2). Microorganisms that oxidize ammonia are found everywhere in the environment, including soils, freshwater and marine environments, artificial ecosystems like WWTPs, and even human skin.[8] The number and variety of ammonia oxidizers vary greatly between habitats, which is significant for the proper functioning of the ecosystem since various ammonia oxidizers react to environmental shifts in different ways.

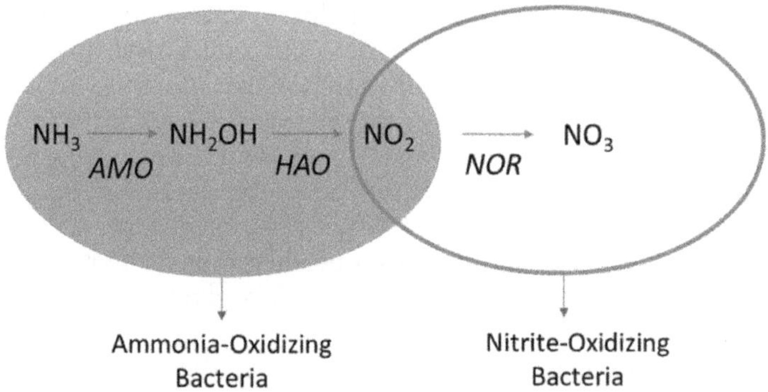

Figure 11.1 Ammonia oxidizing bacteria convert ammonia to nitrite using ammonia monooxygenase (AMO) and hydroxylamine oxidoreductase (HAO). Nitrite oxidizers use the enzyme nitrite oxidoreductase (NOR) for converting nitrite to nitrate.

The enzyme initiating the early oxidation of NH_3, ammonia monooxygenase (AMO), has become one of the most extensively utilized molecular indicators in environmental microbiology.[9] *Nitrosomonas* and *Nitrosospira* are well-known β-ammonia oxidizers,[10] whereas *Nitrosococcus* is a γ-proteobacterial genus,[11] but this doesn't include *Nitrosomonas mobilis*, a related *Nitrosomonas*,[12] as shown in Figure 11.1. Various conditions that define the biological niches of various ammonia oxidizers and the factors that lead to environmental nitrogen fluxes have been thoroughly investigated. The two most constant drivers of ammonia oxidizer variety and abundance are environmental pH and ammonia concentration.[13] Additional variables that determine the dispersion of ammonia-oxidizing microbes in the environment include their diverse sensitivities to organic chemicals, metals, temperature, salinity, and light suppression.[14]

11.2.2 Physiology of Ammonium Oxidizers

Ammonium oxidizers can be classified into three groups, namely ammonia-oxidizing bacteria (AOB), ammonia-oxidizing archaea (AOA), and comammox *Nitrospira*. They are autotrophic but their inorganic carbon absorption routes differ. AOB have lower substrate affinities and greater half-saturation constants (K_m) than AOA and comammox *Nitrospira*.[15] When the substrate concentration is low, greater affinity is favourable, so AOA are more abundant in the ocean and low nitrogen soils, but AOB dominate in WWTPs containing higher ammonium levels.[16] The Calvin cycle and the reductive tricarboxylic acid cycle are used by both AOB and comammox *Nitrospira* to fix ambient CO_2.[17] The large membranes observed in AOB have long been thought to be there to accommodate as much ammonia monooxygenase as possible. Both aerobic and anaerobic ammonia oxidation provides energy for proteobacterial ammonia oxidizers to proliferate.[18]

11.2.3 Ecological Niche of N Oxidizers

Nitrite oxidizing bacteria, such as bacteria of the genera *Nitrobacter*, *Nitrococcus*, and *Nitrospira*,[19] help in the oxidation of nitrite to nitrate. Sequences from *Candidatus Nitrotoga*[20] have now been identified in wastewater treatment facilities (WWTPs) and activated sludge. The *Nitrobacter* and *Nitrococcus* genera are alpha-proteobacteria, whereas *Nitrospira* is evolutionarily distinct from any other cultured species and belongs to a separate division.[21] The other genera are more adaptable, like facultative autotrophs and anaerobes that can thrive on heterotrophic substrates like pyruvate and can even perform the first phase of denitrification.[22]

11.2.4 Physiology of N Oxidizers

As compared to the ammonia oxidizers, nitrite oxidizers are more flexible.[23] Studies have shown the efficiency of hydroxylamine in completely inhibiting nitrite oxidation to nitrate at dosages of 7.0 and 8.9 $mg\,L^{-1}$ for pH values of 8.4 and 7.6, respectively. The nitrite oxidoreductase, oxidizing nitrite to create nitrate, is a crucial enzyme in nitrite-oxidizing bacteria.[24] Organisms that oxidize nitrite are usually lithoautotrophic.[25] When cells develop in a mixotrophic environment, they grow at a faster pace. Several *Nitrobacter* strains may thrive heterotrophically in both oxic and anoxic environments.

11.3 Process of N-removal

There are a range of physicochemical and biological techniques that can remove ammonium from wastewater. Biological nitrogen removal has been widely used in place of physicochemical techniques since it is efficient and affordable. However, there are several drawbacks of the conventional N-removal process, which has resulted in the emergence of innovative modern processes such as the SHARON process, the CANON process, and the ANAMMOX process. The ANAMMOX procedures are used in conjunction with partial-autotrophic-nitritation, which involves ammonium oxidizing bacteria (AOB) converting about half of the ammonium to nitrite.[26]

11.3.1 Conventional N-removal Process

Conventional microbial nitrogen removal includes two steps: autotrophic nitrification and heterotrophic denitrification. Aerobic nitrification is the conversion of ammonium (NH_4^+) to nitrite (NO_2^-) and further to nitrate (NO_3^-) with molecular oxygen as an electron acceptor. Heterotrophic denitrification, also known as anoxic denitrification, involves the conversion of nitrite (NO_2^-) and nitrate (NO_3^-) to gaseous nitrogen with methanol, acetate, ethanol, lactate and glucose as electron donors.

The conventional aerobic nitrification process requires a large amount of oxygen whereas during the denitrification, a considerable amount of organic carbon is required. For instance, 4.2 g of oxygen is required for each gram of ammonium to nitrite conversion and 2.47 g of methanol is utilized for each gram of nitrite to nitrate complete denitrification.[27] Full-scale denitrification is economically unfeasible due to the need for additional electron donors such as methanol.

Methanotrophs oxidize methane to carbon dioxide and water and produce intermediate compounds under favourable environmental conditions that help in the anoxic denitrification as the carbon source.[28,29] However, the process is quite slow. Moreover, for the complete removal of nitrogen from wastewater, a considerable amount of carbon is necessary, which is limited in nature.[30]

11.3.2 SHARON Process

The SHARON (single reactor system for high ammonia removal over nitrite) technology is a revolutionary biological nitrogen removal technique. In this process, nitrite acts as the intermediate for nitrification and denitrification.[30] During nitrification, ammonium is oxidized to hydroxylamine, which is then oxidized to nitrite. Nitrite is then converted to nitrate. During denitrification, nitrate is reduced to nitrite which is converted to gaseous NO, N_2O and N_2. This method involves partial nitrification of ammonium to nitrate, which lowers the cost of aeration significantly.[31] The operational factors are regulated in a chemostat operation to achieve steady partial nitrification.[31] It is possible to attain a nitrogen removal effectiveness of 90%. The SHARON method requires 25% less aeration energy and 40% less additional carbon than classical nitrification and denitrification using nitrate.[30]

11.3.3 ANAMMOX Process

Denitrification of nitrite using ammonia as the electron donor is known as the anammox process. Researchers have found that ammonium can be oxidized by ammonium oxidizers in anoxic circumstances. For example, in a lab-scale anaerobic fluidized denitrifying bed used to handle the effluent from a methanogenic reactor,[32] and for anaerobic ammonium oxidation (ANAMMOX)[33] as shown in Figure 11.2. The major carbon source for the development of ANAMMOX bacteria is carbon dioxide. This technique involves the oxidation of ammonium into hydrazine *via* hydroxylamine. N_2H_4 derived reducing equivalents are reduced to hydroxylamine and nitrogen gas.[34] This method holds a lot of promise for removing ammonium from sludge digesting effluent.

The process requires a low operation cost due to its low biomass production. However, because of its low biomass yield, an effective sludge retention system is necessary, and an extensive start-up duration is required to achieve an adequate biomass concentration.[35–37]

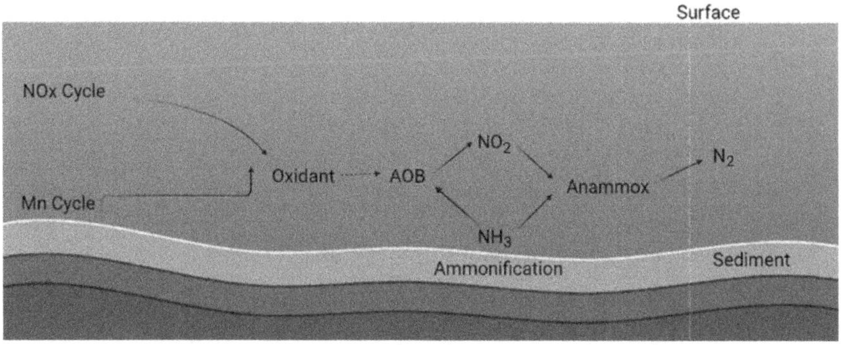

Figure 11.2 Ammonia is released in the anaerobic sediments through ammonification of organic matter.

11.3.4 CANON Process

Dijkman and Strous (1999) discussed the 'Completely autotrophic nitrogen removal over nitrite (CANON)' method for entirely autotrophic nitrogen removal over nitrite as a novel biological nitrogen removal mechanism. This method eliminates ammonium from wastewater that contains just a small quantity of organic matter. Partial nitrification and anoxic oxidation of ammonia (ANAMMOX) are used in this procedure.[38] Under oxygen-limited circumstances, the system depends on a steady relationship between two autotrophic microbe populations.[39] Autotrophic microorganisms use nitrate as an intermediary to convert ammonia to dinitrogen gas. The CANON method is a cost-effective and effective wastewater treatment technique. Moreover, the process reduces a lot of space and energy. According to Kuai and Verstraete, 1998, compared to a typical nitrogen removal method, the autotrophic process uses 63% less oxygen and 100% less reducing compounds.[39]

11.4 Emerging Strategies Manoeuvred for Enhanced Accessibility of Anammox

Substrate concentrations have a significant impact on bacterial growth and metabolism, and consequently community composition as reported by Polz and Cordero, 2016. Under substrate-limited conditions, timely control of gene expression is necessary to tolerate harsh environments and boost bacterial multiplication. Microorganisms prefer to express enzymes and transporters in catabolic pathways as much as possible at low substrate concentrations, even if they don't require that much metabolism at that moment. Moreover, metabolic redundancies of this type appear to promote increased interspecific interactions and dependence.

New technologies for the long-term removal of ammonium from wastewater are required by stringent nitrogen discharge limits. One of these systems is based on the anaerobic ammonium oxidation (anammox) process,

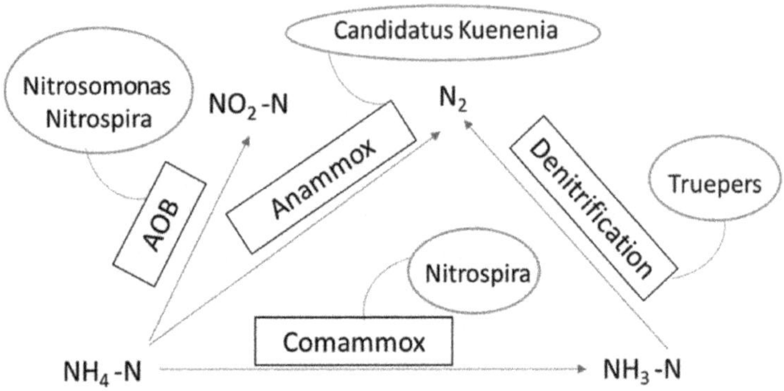

Figure 11.3 Nitrogen removal by partial nitrification anammox and comammox.

which is a new and potent technique for treating high-nitrogen wastewaters[40,41] as shown in Figure 11.3.

According to recent research, lithotrophic ammonia oxidizers can denitrify in both oxygen-limited and anoxic environments. These bacteria may also oxidize ammonia using N_2O_4 (a dimeric form of NO_2) as an oxidant in both oxic and anoxic settings.

Anammox and Nitrosomonas (Nitrobacteraceae), two genera of lithoautotrophic ammonia oxidizers, are considered to be natural companions in habitats with low oxygen availability. Initially identified in the 1980s, the use of anaerobic ammonium oxidation (anammox) in a variety of water technologies has grown in popularity. The partial nitritation and anammox (PN/A) method has environmental benefits over traditional nitrogen removal, such as a 60% decrease in oxygen consumption, a 95% reduction in biodegradable carbon, and an 80% decrease in sludge, which are the major objectives of current wastewater treatment.[42,43]

11.5 Application of Anammox in Wastewater Management

The anammox technology has opened up new opportunities in the wastewater treatment industry. In large-scale treatment plants, this technique has the potential to replace traditional nitrification/denitrification stages. In wastewater systems, the anammox process has resulted in greater nitrogen removal rates and lower energy needs than traditional nitrogen removal methods.[44] Anammox bacteria are useful because of their high salt tolerance, wide temperature tolerance, and capacity to remove organic material.[45] ANAMMOX brings down the energy consumption by 63% and lowers the CO_2 production by 90%[46] because of its unique metabolic ability to combine ammonium and nitrite or nitrate to form nitrogen gas without consuming organic carbon.[46] The property of partial nitritation shown by ANAMMOX is the reason behind its popularity because of its low capital costs and simple operation.[47]

11.6 Strategies to Improve Efficacy of Anammox in Wastewater Management

During the operation of an ANAMMOX system, several parameters have a direct impact on nitrogen removal performance. Temperature, pH, DO, nitrogen loading, and carbon source content are the most important to properly manage in order to generate a consistent effluent that meets discharge standards. The emergence of some unusual harmful organic and inorganic chemicals as a result of environmental catastrophes or accidents, and the difficulty of industrial wastewater, can impair or obstruct anammox systems' normal operation.[48] Anammox performance is restorable by carefully monitoring its activity and microbial dynamic changes in the system.[49]

11.6.1 Temperature

The ANAMMOX process is effective at high temperatures, but low temperatures seen in the winter months pose a barrier to commercializing this wastewater treatment technique. Many studies showed full-scale sidestream anammox systems for treating digested sewage, landfill leachate, and rejected water.[50,51] They discovered that high temperatures, generally greater than 30 °C, were necessary for high ANAMMOX activity and nitrogen removal efficiency in the sidestream anammox process.[51] Because the ANAMMOX process has previously demonstrated strong performance at low temperatures, it would increase the likelihood of widespread applicability of ANAMMOX.[51]

11.6.2 pH

Maintaining pH is critical to sustaining high nitrogen removal over an extended length of time.[52] The favourable range of pH that is essential for the optimum efficiency of nitrogen removal is 6.7–8.3.[53,54] The anammox bacterial membrane's poor permeability and protons' restricted diffusion capacity can protect the bacterium against alkaline or acid environments. When the pH of the enrichment culture was regulated from 6.8–7.0, *Brocadia anammoxidans* and "*Candidatus Anammoxoglobus propionicus*" were found to be prominent in the aerobic granules and leachate sludge, respectively.[55] However, at pH 7.8–8, "*Candidatus Brocadia anammoxidans*" and *Kuenenia stuttgartiensis* were found to be predominant.[56]

11.6.3 Dissolved Oxygen (DO)

The most effective strategy for maintaining nitrite and ammonium levels for the anammox reaction and reducing NOB activity is intermittent aeration with instantaneous DO regulation.[57] Low oxygen concentrations or anoxic circumstances promote anammox metabolism, especially enzyme synthesis, and trigger the induction of the anammox process.[58] According to Liu *et al.*,[59] a DO level of less than 0.04 mg L^{-1} can remove a considerable amount of nitrogen in anammox activity.

11.6.4 Nitrogen Loading

High nitrogen concentrations are frequent in real wastewaters. As a result, using an acceptable loading rate and effluent recirculation to keep influent nitrogen below the inhibition threshold would enhance and maintain the nitrogen removal rate in anammox systems.[60] Lower nitrogen levels are taken into consideration for faster activation and faster growth of anammox.[61] The production of numerous enzymes is inhibited by high salinity, and cell metabolism is slowed by plasmolysis.[62–64]

11.6.5 Carbon Content

The presence of significant quantities of organic carbon in sludge waste decreases the efficacy of anammox activity in various ways.[63] Due to competition between anammox bacteria and heterotrophic denitrifying bacteria, the presence of significant organic matter in the wastewater treatment system has a detrimental impact on ammonium removal by the anammox process. Because denitrifiers need oxygen to digest organic carbon, anammox bacteria can profit in this situation. Organic carbon compounds can be consumed by "*Candidatus Brocadia fulgida*".[65]

11.7 Novel Technologies for Removal of Wastewater Using Nitrogen Removing Bacteria

Nitrogen is an essential nutrient for the survival of living organisms. In the atmosphere, a large proportion of nitrogen exists in N_2 gas form. It can also be converted to nitrate, nitrite, ammonium, and ammonia. The human population is growing day by day and so the dependence on food is rising at an alarming rate. To mitigate this challenge, modern agriculture is producing the required products but at the same time the application of fertilizers has resulted in reactive nitrogen in the environment, causing pollution. The recovery of the reactive nitrogen from wastewater for food production would help to lower energy consumption and cause less pollution.[66] Two techniques can be employed to reduce nitrogen deposition, which include bio-electrochemical systems and photosynthetic microorganisms.

11.7.1 Recovery by Bio-electrochemical Systems

Biological nitrogen recovery using bioelectrochemical systems is based on ammonia/ammonium recovery.[67] In this method, the biological oxidation of organic matter in wastewater at the anodic compartment contributes to energy recovery. Ammonium was recovered from urine at a rate of 9.7 gN per m^2 per day with an energy yield of -10 kJ gN^{-1} at a current density of 2.6 A m^{-2}.[68] Transport of NH_4^+ makes the nitrogen recovery essential and pH gradients of other cations than H^+ are liable to be lost in cation exchange membranes, which has an impact on the performance of bioelectrochemical reactors.[69]

11.7.2 Recovery by Microalgae and Cyanobacteria

There are unicellular species of microscopic photosynthetic microorganisms, microalgae, and cyanobacteria in both freshwater and marine systems. Recently, microalgae and cyanobacteria have been considered alternative systems for biological wastewater treatment with a variety of applications.[70] Nitrate and ammonia can be converted to biomass by microalgae and cyanobacteria rather than being released into the atmosphere as N_2 gas by dissimilatory nitrate reduction.[71] It can then be incorporated into amino acids and proteins, or it can be excreted into the environment.[72]

11.8 Futuristic Goals

The removal of nitrogen from wastewater treatment (WWT) is a critical aim because it creates trouble in the natural habitat by causing damage to water bodies and marine biota.[73]

An improved biomass storage time is an important component that should be enhanced in the future. Numerous methodologies, such as sequential batch reactors, have been utilized in this regard, with positive results.

Numerous techniques, like sequential batch reactors, have yielded good results, or many application areas of sequential-batch anammox systems have been tested at large scale exhibiting great results.[74] Since some pilot studies have reported its potential deployment, further study is needed to verify the suitability of anammox process applications at a wide scale. According to various publications, fixed biofilm anammox systems are even more tolerant to pollutants than traditional suspended biomass systems. The nutrient enrichment of several habitats in recent years has sparked a renewed focus on nitrogen transformation ecology. The usage of amoA has opened up opportunities for assessing AOB variety and size. AOB consists of a variety of protein-coding targets that could be used to characterise AOB groups in samples collected.[75]

A variety of unique strategies are subsequently often used to link function with phylogenetic identification. To detect species that assimilate tagged substrates, micro-autoradiography can be combined with using phylogenetic probes. This method was used to investigate bacteria found in activated sludge.

11.9 Conclusion and Summary

The ecology and genetics of ammonia oxidizers have been thoroughly dissected in recent years, thanks to the significant invention of new species participating in ammonia oxidation. Here, we found that the amount and types of ammonia oxidizers vary significantly among ecosystems, which is important for ecological systems since different ammonia oxidizers adapt to external development in various methods. The amoA gene and the 16S rRNA gene have provided practically everything known about ammonia oxidizer ecology. Ammonia-oxidizers are a type of chemolithotroph that have developed specialised pathways of electron carriers to develop proton motive force from ammonia as its primary source of energy, as well as strategies to withstand cytotoxic metabolic products. The development and culture of novel microbes, the rapid multiplication of genome sequence and functional 'omics evidence, rigorous biochemical research that not only improved the activity of known enzymes and encouraged us to find new enzymes, and environmental research that showed interactions of ammonia-oxidizers to one another and to partner microbes have all expedited our knowledge of ammonia-oxidizing microorganisms. In the wastewater treatment business, the Anammox technology has opened up new avenues. The pH range required for optimal nitrogen removal effectiveness is 6.7–8.3. Low oxygen levels or anoxic conditions encourage anammox metabolism, particularly enzyme synthesis, and initiate the anammox process. The anodic compartment's biological oxidation of organic materials in wastewater contributes to energy recovery. Microalgae and cyanobacteria-based biological remedies may be able to overcome the limits of current tertiary treatment approaches. In the coming years, anammox technologies will have to overcome a number of critical challenges in addition to the possibility of using anammox systems for municipal wastewater treatment.

References

1. M. G. Klotz, *Research on Nitrification and Related Processes, Part B*, Academic Press, 2011.
2. K. Isaka, T. Sumino and S. Tsuneda, High nitrogen removal performance at moderately low temperature utilizing anaerobic ammonium oxidation reactions, *J. Biosci. Bioeng.*, 2007, **103**(5), 486–490.
3. Z. Hu, T. Lotti, M. van Loosdrecht and B. Kartal, Nitrogen removal with the anaerobic ammonium oxidation process, *Biotechnol. Lett.*, 2013, **35**(8), 1145–1154.
4. T. Sumino, K. Isaka, H. Ikuta, Y. Saiki and T. Yokota, Nitrogen removal from wastewater using simultaneous nitrate reduction and anaerobic ammonium oxidation in single reactor, *J. Biosci. Bioeng.*, 2006, **102**(4), 346–351.
5. B. K. Nahak, S. Preetam, D. Sharma, S. K. Shukla, M. Syväjärvi, D. C. Toncu and A. Tiwari, Advancements in net-zero pertinency of lignocellulosic

biomass for climate neutral energy production, *Renewable Sustainable Energy Rev.*, 2022, **161**, 112393.
6. I. Schmidt, O. Sliekers, M. Schmid, E. Bock, J. Fuerst, J. G. Kuenen and M. Strous, New concepts of microbial treatment processes for the nitrogen removal in wastewater, *FEMS Microbiol. Rev.*, 2003, **27**(4), 481–492.
7. A. González-Martínez, K. Calderón and J. González-López, New concepts of microbial treatment processes for the nitrogen removal: effect of protein and amino acids degradation, *Amino Acids*, 2016, **48**(5), 1123–1130.
8. M. Mußmann, I. Brito, A. Pitcher, J. S. S. Damsté, R. Hatzenpichler, A. Richter and I. M. Head, Thaumarchaeotes abundant in refinery nitrifying sludges express amoA but are not obligate autotrophic ammonia oxidizers, *Proc. Natl. Acad. Sci. U. S. A.*, 2011, **108**(40), 16771–16776.
9. A. Bateman, The structure of a domain common to archaebacteria and the homocystinuria disease protein, *Trends Biochem. Sci.*, 1997, **22**(1), 12–13.
10. R. Hatzenpichler, Diversity, physiology, and niche differentiation of ammonia-oxidizing archaea, *Appl. Environ. Microbiol.*, 2012, **78**(21), 7501–7510.
11. H. Daims, E. V. Lebedeva, P. Pjevac, P. Han, C. Herbold, M. Albertsen and M. Wagner, Complete nitrification by Nitrospira bacteria, *Nature*, 2015, **528**(7583), 504–509.
12. C. E. Lawson and S. Lücker, Complete ammonia oxidation: an important control on nitrification in engineered ecosystems?, *Curr. Opin. Biotechnol.*, 2018, **50**, 158–165.
13. L. E. Lehtovirta-Morley, Ammonia oxidation: ecology, physiology, biochemistry and why they must all come together, *FEMS Microbiol. Lett.*, 2018, **365**(9), fny058.
14. S. N. Merbt, D. A. Stahl, E. O. Casamayor, E. Martí, G. W. Nicol and J. I. Prosser, Differential photoinhibition of bacterial and archaeal ammonia oxidation, *FEMS Microbiol. Lett.*, 2012, **327**(1), 41–46.
15. J. I. Prosser and G. W. Nicol, Archaeal and bacterial ammonia-oxidisers in soil: the quest for niche specialisation and differentiation, *Trends Microbiol.*, 2012, **20**(11), 523–531.
16. W. Martens-Habbena, P. M. Berube, H. Urakawa, J. R. de La Torre and D. A. Stahl, Ammonia oxidation kinetics determine niche separation of nitrifying Archaea and Bacteria, *Nature*, 2009, **461**(7266), 976–979.
17. I. A. Berg, D. Kockelkorn, W. Buckel and G. Fuchs, A 3-hydroxypropionate/4-hydroxybutyrate autotrophic carbon dioxide assimilation pathway in Archaea, *Science*, 2007, **318**(5857), 1782–1786.
18. M. Könneke, D. M. Schubert, P. C. Brown, M. Hügler, S. Standfest, T. Schwander and I. A. Berg, Ammonia-oxidizing archaea use the most energy-efficient aerobic pathway for CO_2 fixation, *Proc. Natl. Acad. Sci. U. S. A.*, 2014, **111**(22), 8239–8244.
19. J. I. Prosser, Autotrophic nitrification in bacteria, *Adv. Microb. Physiol.*, 1990, **30**, 125–181.

20. A. I. Al-Alawi, N. Y. Al-Marzooqi and Y. F. Mohammed, Organizational culture and knowledge sharing: critical success factors, *J. knowl. Manage.*, 2007, **11**, 22–42.
21. S. Ehrich, D. Behrens, E. Lebedeva, W. Ludwig and E. Bock, A new obligately chemolithoautotrophic, nitrite-oxidizing bacterium, Nitrospira moscoviensis sp. nov. and its phylogenetic relationship, *Arch. Microbiol.*, 1995, **164**(1), 16–23.
22. E. Bock, H. P. Koops, H. Harms and B. Ahlers, The biochemistry of nitrifying organisms, *Variations in autotrophic life*, 1991, vol. 15, pp. 171–200.
23. I. Schmidt, O. Sliekers, M. Schmid, E. Bock, J. Fuerst, J. G. Kuenen and M. Strous, New concepts of microbial treatment processes for the nitrogen removal in wastewater, *FEMS Microbiol. Rev.*, 2003, **27**(4), 481–492.
24. Y. Tanaka, Y. Fukumori and T. Yamanaka, Purification of cytochrome a 1 c 1 from Nitrobacter agilis and characterization of nitrite oxidation system of the bacterium, *Arch. Microbiol.*, 1983, **135**(4), 265–271.
25. D. Y. Sorokin, G. Muyzer, T. Brinkhoff, J. Gijs Kuenen and M. S. Jetten, Isolation and characterization of a novel facultatively alkaliphilic Nitrobacter species, N. alkalicus sp. nov, *Arch. Microbiol.*, 1998, **170**(5), 345–352.
26. P. L. McCarty, Biological denitrification of wastewaters by addition of organic materials, in *Proceedings of the 24th Industrial Waste Conference*, Purdue University Libraries, 1969.
27. W. Gujer and D. A. Jenkins, *The Contact Stabilization Process: Oxygen and Nitrogen Mass Balances, No. 74*. Sanitary Engineering Research Laboratory, College of Engineering and School of Public Health, University of California, 1974.
28. C. Costa, C. Dijkema, M. Friedrich, P. Garcia-Encina, F. Fernandez-Polanco and A. J. M. Stams, Denitrification with methane as electron donor in oxygen-limited bioreactors, *Appl. Microbiol. Biotechnol.*, 2000, **53**(6), 754–762.
29. H. J. Lee, J. H. Bae and K. M. Cho, Simultaneous nitrification and denitrification in a mixed methanotrophic culture, *Biotechnol. Lett.*, 2001, **23**(12), 935–941.
30. U. G. J. M. Van Dongen, M. S. Jetten and M. van Loosdrecht, The SHARON®-Anammox® process for treatment of ammonium rich wastewater, *Water Sci. Technol.*, 2001, **44**(1), 153–160.
31. T. Khin and A. P. Annachhatre, Novel microbial nitrogen removal processes, *Biotechnol. Adv.*, 2004, **22**(7), 519–532.
32. J. Fu, Q. Zhang, B. Huang, N. Fan and R. Jin, A review on anammox process for the treatment of antibiotic-containing wastewater: Linking effects with corresponding mechanisms, *Front. Environ. Sci. Eng.*, 2021, **15**(1), 1–15.
33. K. Nera, J. Jetten, M. Biddlestone and O. Klein, 'Who wants to silence us'? Perceived discrimination of conspiracy theory believers increases

'conspiracy theorist' identification when it comes from powerholders–But not from the general public, *Br. J. Soc. Psychol.*, 2022, **61**(4), 1263–1285.
34. G. Berkers, P. van Mourik, A. M. Vonk, E. Kruisselbrink, J. F. Dekkers, K. M. de Winter-de Groot and C. K. van der Ent, Rectal organoids enable personalized treatment of cystic fibrosis, *Cell Rep.*, 2019, **26**(7), 1701–1708.
35. Q. Zhang, J. G. Lin, Z. Kong and Y. Zhang, A critical review of exogenous additives for improving the anammox process, *Sci. Total Environ.*, 2022, 155074.
36. Z. C. Zhao, G. J. Xie, B. F. Liu, D. F. Xing, J. Ding, H. J. Han and N. Q. Ren, A review of quorum sensing improving partial nitritation-anammox process: Functions, mechanisms and prospects, *Sci. Total Environ.*, 2021, **765**, 142703.
37. D. Sobotka, J. Zhai and J. Makinia, Generalized temperature dependence model for anammox process kinetics, *Sci. Total Environ.*, 2021, **775**, 145760.
38. K. A. Third, A. O. Sliekers, J. G. Kuenen and M. S. Jetten, The CANON system (Completely Autotrophic Nitrogen-removal Over Nitrite) under ammonium limitation: interaction and competition between three groups of bacteria, *Syst. Appl. Microbiol.*, 2001, **24**(4), 588–596.
39. A. Tawfik and M. Elsamadony, Completely autotrophic nitrogen removal over nitrite (CANON) process for polishing of anaerobic effluent containing ammonia, in *Post Treatments of Anaerobically Treated Effluents*, 2019.
40. X. Li, M. Shi, M. Zhang, W. Li, P. L. Xu, Y. Wang and Y. Huang, Progresses and challenges in sulfur autotrophic denitrification-enhanced Anammox for low carbon and efficient nitrogen removal, *Crit. Rev. Environ. Sci. Technol.*, 2022, 1–16.
41. N. C. Silveira, G. H. D. Oliveira, M. H. R. Z. Damianovic and E. Foresti, Two-stage partial nitrification-Anammox process for nitrogen removal from slaughterhouse wastewater: Evaluation of the nitrogen loading rate and microbial community analysis, *J. Environ. Manage.*, 2021, **296**, 113214.
42. U. Ghimire, G. Sarpong and V. G. Gude, Transitioning wastewater treatment plants toward circular economy and energy sustainability, *ACS Omega*, 2021, **6**(18), 11794–11803.
43. M. Laureni, D. G. Weissbrodt, K. Villez, O. Robin, N. De Jonge, A. Rosenthal and A. Joss, Biomass segregation between biofilm and flocs improves the control of nitrite-oxidizing bacteria in mainstream partial nitritation and anammox processes, *Water Res.*, 2019, **154**, 104–116.
44. K. Liu, X. Luo, J. J. Jiao, J. D. Gu and R. Aravena, Gene abundances of AOA, AOB, and anammox controlled by groundwater chemistry of the Pearl River Delta, China, *China Geol.*, 2021, **4**, 463–475.
45. L. Wang, W. Gu, Y. Liu, P. Liang, X. Zhang and X. Huang, Challenges, solutions and prospects of mainstream anammox-based process for municipal wastewater treatment, *Sci. Total Environ.*, 2022, 153351.

46. S. Q. Ni, H. A. Ahmad, Y. Zhao, Q. Li, Y. Dong, S. Ahmad and Z. Qiao, Energy-efficient anaerobic ammonia removal: from laboratory to full-scale application, in *Microbial Bioremediation & Biodegradation*, Springer, Singapore, 2020, pp. 505–526.
47. C. Reino, M. E. Suárez-Ojeda, J. Pérez and J. Carrera, Stable long-term operation of an upflow anammox sludge bed reactor at mainstream conditions, *Water Res.*, 2018, **128**, 331–340.
48. P. N. Egbuikwem, G. C. Obiechefu, F. I. Hai, M. C. E. Devanadera and D. P. Saroj, Potential of suspended growth biological processes for mixed wastewater reclamation and reuse in agriculture: challenges and opportunities, *Environ. Technol. Rev.*, 2021, **10**(1), 77–110.
49. P. N. Egbuikwem, G. C. Obiechefu, F. I. Hai, M. C. E. Devanadera and D. P. Saroj, Potential of suspended growth biological processes for mixed wastewater reclamation and reuse in agriculture: challenges and opportunities, *Environm. Technol. Rev.*, 2021, **10**(1), 77–110.
50. M. Strous, E. Van Gerven, P. Zheng, J. G. Kuenen and M. S. Jetten, Ammonium removal from concentrated waste streams with the anaerobic ammonium oxidation (anammox) process in different reactor configurations, *Water Res.*, 1997, **31**(8), 1955–1962.
51. B. Kartal, J. V. Kuenen and M. C. M. Van Loosdrecht, Sewage treatment with anammox, *Science*, 2010, **328**(5979), 702–703.
52. M. Tomaszewski, G. Cema and A. Ziembińska-Buczyńska, Influence of temperature and pH on the anammox process: a review and meta-analysis, *Chemosphere*, 2017, **182**, 203–214.
53. J. Yang, L. Zhang, D. Hira, Y. Fukuzaki and K. Furukawa, Anammox treatment of high-salinity wastewater at ambient temperature, *Bioresour. Technol.*, 2011, **102**(3), 2367–2372.
54. J. Yang, Nitrogen removal from high salinity wastewater by Anammox process at ambient temperature, *Bioresour. Technol.*, 2011, **102**, 2367–2372.
55. S. Q. Ni, A. Fessehaie, P. H. Lee, B. Y. Gao, X. Xu and S. Sung, Interaction of anammox bacteria and inactive methanogenic granules under high nitrogen selective pressure, *Bioresour. Technol.*, 2010, **101**(18), 6910–6915.
56. S. He, Q. Niu, H. Ma, Y. Zhang and Y. Y. Li, The treatment performance and the bacteria preservation of anammox: a review, *Water, Air, Soil Pollut.*, 2015, **226**(5), 1–16.
57. J. Li, J. Li, R. Gao, M. Wang, L. Yang, X. Wang and Y. Peng, A critical review of one-stage anammox processes for treating industrial wastewater: optimization strategies based on key functional microorganisms, *Bioresour. Technol.*, 2018, **265**, 498–505.
58. T. Dalsgaard, B. Thamdrup, L. Farías and N. P. Revsbech, Anammox and denitrification in the oxygen minimum zone of the eastern South Pacific, *Limnol. Oceanogr.*, 2012, **57**(5), 1331–1346.
59. S. Liu, F. Yang, Y. Xue, Z. Gong, H. Chen, T. Wang and Z. Su, Evaluation of oxygen adaptation and identification of functional bacteria

composition for anammox consortium in non-woven biological rotating contactor, *Bioresour. Technol.*, 2008, **99**(17), 8273–8279.
60. M. Strous, J. A. Fuerst, E. H. Kramer, S. Logemann, G. Muyzer, K. T. van de Pas-Schoonen and M. S. Jetten, Missing lithotroph identified as new planctomycete, *Nature*, 1999, **400**(6743), 446–449.
61. J. Y. Jung, S. H. Kang, Y. C. Chung and D. H. Ahn, Factors affecting the activity of anammox bacteria during start up in the continuous culture reactor, *Water Sci. Technol.*, 2007, **55**(1–2), 459–468.
62. R. C. Jin, G. F. Yang, J. J. Yu and P. Zheng, The inhibition of the Anammox process: a review, *Chem. Eng. J.*, 2012, **197**, 67–79.
63. C. J. Tang, P. Zheng, Q. Mahmood and J. W. Chen, Start-up and inhibition analysis of the Anammox process seeded with anaerobic granular sludge, *J. Ind. Microbiol. Biotechnol.*, 2009, **36**(8), 1093.
64. B. Kartal, M. M. Kuypers, G. Lavik, J. Schalk, H. J. Op den Camp, M. S. Jetten and M. Strous, Anammox bacteria disguised as denitrifiers: nitrate reduction to dinitrogen gas via nitrite and ammonium, *Environ. Microbiol.*, 2007, **9**(3), 635–642.
65. B. Kartal, J. Rattray, L. A. van Niftrik, J. van de Vossenberg, M. C. Schmid, R. I. Webb and M. Strous, Candidatus "Anammoxoglobus propionicus" a new propionate oxidizing species of anaerobic ammonium oxidizing bacteria, *Syst. Appl. Microbiol.*, 2007, **30**(1), 39–49.
66. S. Matassa, D. J. Batstone, T. Hülsen, J. Schnoor and W. Verstraete, Can direct conversion of used nitrogen to new feed and protein help feed the world?, *Environm. Sci. Technol.*, 2015, **49**(9), 5247–5254.
67. P. Kuntke, T. H. J. A. Sleutels, M. Rodríguez Arredondo, S. Georg, S. G. Barbosa, A. Ter Heijne and C. J. N. Buisman, (Bio) electrochemical ammonia recovery: progress and perspectives, *Appl. Microbiol. Biotechnol.*, 2018, **102**(9), 3865–3878.
68. P. Kuntke, Nutrient and energy recovery from urine, PhD thesis, Wageningen University and Research, 2013.
69. D. Cecconet, S. Zou, A. G. Capodaglio and Z. He, Evaluation of energy consumption of treating nitrate-contaminated groundwater by bioelectrochemical systems, *Sci. Total Environ.*, 2018, **636**, 881–890.
70. N. Moondra, N. D. Jariwala and R. A. Christian, Sustainable treatment of domestic wastewater through microalgae, *Int. J. Phytorem.*, 2020, **22**(14), 1480–1486.
71. S. Matassa, D. J. Batstone, T. Hülsen, J. Schnoor and W. Verstraete, Can direct conversion of used nitrogen to new feed and protein help feed the world?, *Environ. Sci. Technol.*, 2015, **49**(9), 5247–5254.
72. V. Evangelista, A. M. Frassanito, V. Passarelli, L. Barsanti and P. Gualtieri, Microspectroscopy of the photosynthetic compartment of algae, *Photochem. Photobiol.*, 2006, **82**(4), 1039–1046.
73. A. Gonzalez-Martinez, B. Muñoz-Palazon, A. Rodriguez-Sanchez and J. Gonzalez-Lopez, New concepts in anammox processes for wastewater nitrogen removal: recent advances and future prospects, *FEMS Microbiol. Lett.*, 2018, **365**(6), fny031.

74. A. Gonzalez-Martinez, B. Muñoz-Palazon, A. Rodriguez-Sanchez, P. Maza-Márquez, A. Mikola, J. Gonzalez-Lopez and R. Vahala, Start-up and operation of an aerobic granular sludge system under low working temperature inoculated with cold-adapted activated sludge from Finland, *Bioresour. Technol.*, 2017, **239**, 180–189.
75. D. J. Miller and D. J. D. Nicholas, Characterization of a soluble cytochrome oxidase/nitrite reductase from Nitrosomonas europaea, *Microbiology*, 1985, **131**(10), 2851–2854.

Subject Index